An Introduction to Biotechnology

An Introduction to Biotechnology

Edited by
Maison Owen

Larsen & Keller
www.larsen-keller.com

An Introduction to Biotechnology
Edited by Maison Owen
ISBN: 978-1-63549-050-3 (Hardback)

⊟ Larsen & Keller

Published by Larsen and Keller Education,
5 Penn Plaza,
19th Floor,
New York, NY 10001, USA

Cataloging-in-Publication Data

An introduction to biotechnology / edited by Maison Owen.
    p. cm.
Includes bibliographical references and index.
ISBN 978-1-63549-050-3
1. Biotechnology. 2. Genetic engineering.
I. Owen, Maison.
TP248.2 .I58 2017
660.6--dc23

The publisher's policy is to use permanent paper from mills that operate a sustainable forestry policy. Furthermore, the publisher ensures that the text paper and cover boards used have met acceptable environmental accreditation standards.

Printed and bound in the United States of America.

For more information regarding Larsen and Keller Education and its products, please visit the publisher's website www.larsen-keller.com

Table of Contents

Preface

This book provides comprehensive insights into the field of biotechnology. It discusses in detail the various concepts and theories of this field. Biotechnology refers to the science of using biological systems or living systems for the design of new and improved products or processes. This book is a compilation of chapters that discuss the most vital concepts in the field of biotechnology. As this field is emerging at a rapid pace, the contents of this textbook will help the readers understand the modern concepts and applications of the subject. Coherent flow of topics, student-friendly language and extensive use of examples make this book an invaluable source of knowledge.

A short introduction to every chapter is written below to provide an overview of the content of the book:

Chapter 1 - Biotechnology is used to develop and produce technology from living organisms, which can improve our health and the well-being of our planet as well. Modern biotechnology has more challenges to face, as elucidated in the following text. The chapter will provide an integrated understanding of biotechnology; **Chapter 2** - Biobased economy, green revolution, biomimetic or tissue culture are important concepts of biotechnology dealt within this chapter. The following chapter unfolds the crucial aspects in a critical yet systematic manner and elucidates the crucial theories and principles of biotechnology. The major concepts of biotechnology are discussed in this chapter; **Chapter 3** - Selective breeding is the process by which humans breed animals and plants with specific characteristics. This chapter discusses the process of selective breeding, biosynthesis, catalysis and hybrid. It explains in detail the process of selective breeding and hybrid breeding; **Chapter 4** - Biotechnology is best understood in confluence with the major topics listed in the following chapter. The applications of biotechnology covered in this chapter are cloning, genetic engineering, recombinant DNA and tissue engineering. The topics discussed in the chapter are of great importance to broaden the existing knowledge on biotechnology; **Chapter 5** - The products developed from the use of living systems and organisms are known as biotechnological products. Some of the biological products discussed in the chapter are bioSteel, biopolymer, burton, arctic apples and specialty drugs. The content provides an overview of the subject matter incorporating all the major products of biotechnology; **Chapter 6** - Biotechnology is an interdisciplinary subject and spans across varied fields. This chapter covers agriculture, agricultural science, food industry and processing and medicine. The chapter provides a plethora of allied fields of biotechnology for a better comprehension of the respective subject.

I extend my sincere thanks to the publisher for considering me worthy of this task. Finally, I thank my family for being a source of support and help.

Editor

Introduction to Biotechnology

Biotechnology is used to develop and produce technology from living organisms, which can improve our health and the well-being of our planet as well. Modern biotechnology has more challenges to face, as elucidated in the following text. The chapter will provide an integrated understanding of biotechnology.

Biotechnology is the use of living systems and organisms to develop or make products, or "any technological application that uses biological systems, living organisms or derivatives thereof, to make or modify products or processes for specific use" (UN Convention on Biological Diversity, Art. 2). Depending on the tools and applications, it often overlaps with the (related) fields of bioengineering, biomedical engineering, biomanufacturing, etc.

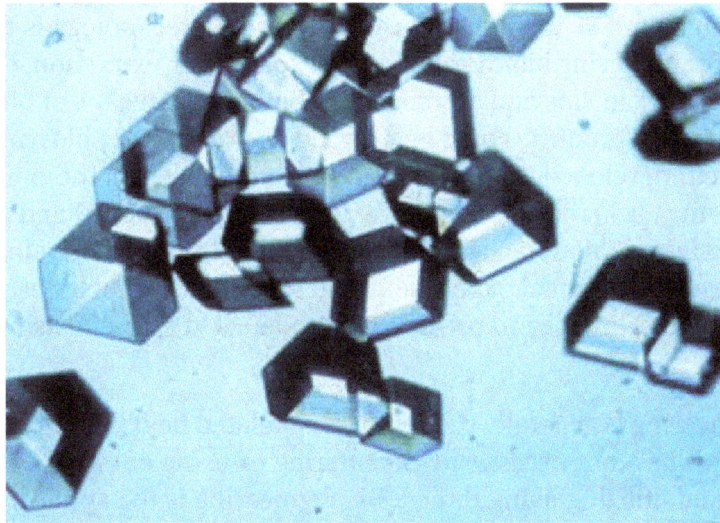

Insulin crystals

For thousands of years, humankind has used biotechnology in agriculture, food production, and medicine. The term is largely believed to have been coined in 1919 by Hungarian engineer Károly Ereky. In the late 20th and early 21st century, biotechnology has expanded to include new and diverse sciences such as genomics, recombinant gene techniques, applied immunology, and development of pharmaceutical therapies and diagnostic tests.

Definitions

The wide concept of "biotech" or "biotechnology" encompasses a wide range of procedures for modifying living organisms according to human purposes, going back to domestication of animals, cultivation of the plants, and "improvements" to these through breeding programs that employ artificial selection and hybridization. Modern usage also includes genetic engineering as well as cell and tissue culture technologies. The American Chemical Society defines biotechnology as the application of biological organisms, systems, or processes by various industries to learning about the science of life and the improvement of the value of materials and organisms such as pharmaceuticals, crops, and livestock. As per European Federation of Biotechnology, Biotechnology is the integration of natural science and organisms, cells, parts thereof, and molecular analogues for products and services. Biotechnology also writes on the pure biological sciences (animal cell culture, biochemistry, cell biology, embryology, genetics, microbiology, and molecular biology). In many instances, it is also dependent on knowledge and methods from outside the sphere of biology including:

- bioinformatics, a new brand of computer science

- bioprocess engineering

- biorobotics

- chemical engineering

Conversely, modern biological sciences (including even concepts such as molecular ecology) are intimately entwined and heavily dependent on the methods developed through biotechnology and what is commonly thought of as the life sciences industry. Biotechnology is the research and development in the laboratory using bioinformatics for exploration, extraction, exploitation and production from any living organisms and any source of biomass by means of biochemical engineering where high value-added products could be planned (reproduced by biosynthesis, for example), forecasted, formulated, developed, manufactured and marketed for the purpose of sustainable operations (for the return from bottomless initial investment on R & D) and gaining durable patents rights (for exclusives rights for sales, and prior to this to receive national and international approval from the results on animal experiment and human experiment, especially on the pharmaceutical branch of biotechnology to prevent any undetected side-effects or safety concerns by using the products).

By contrast, bioengineering is generally thought of as a related field that more heavily emphasizes higher systems approaches (not necessarily the altering or using of biological materials *directly*) for interfacing with and utilizing living things. Bioengineering is the application of the principles of engineering and natural sciences to tissues, cells and molecules. This can be considered as the use of knowledge from working with and manipulating biology to achieve a result that can improve functions in plants and animals. Relatedly, biomedical engineering is an overlapping field that often draws upon and applies *biotechnology* (by various definitions), especially in certain subfields of biomedical and/or chemical engineering such as tissue engineering, biopharmaceutical engineering, and genetic engineering.

History

Although not normally what first comes to mind, many forms of human-derived agriculture clearly fit the broad definition of "utilizing a biotechnological system to make products". Indeed, the cultivation of plants may be viewed as the earliest biotechnological enterprise.

Agriculture has been theorized to have become the dominant way of producing food since the Neolithic Revolution. Through early biotechnology, the earliest farmers selected and bred the best suited crops, having the highest yields, to produce enough food to support a growing population. As crops and fields became increasingly large and difficult to maintain, it was discovered that specific organisms and their by-products could effectively fertilize, restore nitrogen, and control pests. Throughout the history of agriculture, farmers have inadvertently altered the genetics of their crops through introducing them to new environments and breeding them with other plants — one of the first forms of biotechnology.

Brewing was an early application of biotechnology

These processes also were included in early fermentation of beer. These processes were introduced in early Mesopotamia, Egypt, China and India, and still use the same basic biological methods. In brewing, malted grains (containing enzymes) convert starch from grains into sugar and then adding specific yeasts to produce beer. In this process, carbohydrates in the grains were broken down into alcohols such as ethanol. Later other cultures produced the process of lactic acid fermentation which allowed the fermentation and preservation of other forms of food, such as soy sauce. Fermentation was also used in this time period to produce leavened bread. Although the process of fermentation was not fully understood until Louis Pasteur's work in 1857, it is still the first use of biotechnology to convert a food source into another form.

Before the time of Charles Darwin's work and life, animal and plant scientists had already used selective breeding. Darwin added to that body of work with his scientific observations about the ability of science to change species. These accounts contributed to Darwin's theory of natural selection.

For thousands of years, humans have used selective breeding to improve production of crops and livestock to use them for food. In selective breeding, organisms with desirable characteristics are mated to produce offspring with the same characteristics. For example, this technique was used with corn to produce the largest and sweetest crops.

In the early twentieth century scientists gained a greater understanding of microbiology and explored ways of manufacturing specific products. In 1917, Chaim Weizmann first used a pure microbiological culture in an industrial process, that of manufacturing corn starch using *Clostridium acetobutylicum,* to produce acetone, which the United Kingdom desperately needed to manufacture explosives during World War I.

Biotechnology has also led to the development of antibiotics. In 1928, Alexander Fleming discovered the mold *Penicillium.* His work led to the purification of the antibiotic compound formed by the mold by Howard Florey, Ernst Boris Chain and Norman Heatley – to form what we today know as penicillin. In 1940, penicillin became available for medicinal use to treat bacterial infections in humans.

The field of modern biotechnology is generally thought of as having been born in 1971 when Paul Berg's (Stanford) experiments in gene splicing had early success. Herbert W. Boyer (Univ. Calif. at San Francisco) and Stanley N. Cohen (Stanford) significantly advanced the new technology in 1972 by transferring genetic material into a bacterium, such that the imported material would be reproduced. The commercial viability of a biotechnology industry was significantly expanded on June 16, 1980, when the United States Supreme Court ruled that a genetically modified microorganism could be patented in the case of *Diamond v. Chakrabarty.* Indian-born Ananda Chakrabarty, working for General Electric, had modified a bacterium (of the *Pseudomonas* genus) capable of breaking down crude oil, which he proposed to use in treating oil spills. (Chakrabarty's work did not involve gene manipulation but rather the transfer of entire organelles between strains of the *Pseudomonas* bacterium.)

Revenue in the industry is expected to grow by 12.9% in 2008. Another factor influencing the biotechnology sector's success is improved intellectual property rights legislation—and enforcement—worldwide, as well as strengthened demand for medical and pharmaceutical products to cope with an ageing, and ailing, U.S. population.

Rising demand for biofuels is expected to be good news for the biotechnology sector, with the Department of Energy estimating ethanol usage could reduce U.S. petroleum-derived fuel consumption by up to 30% by 2030. The biotechnology sector has allowed the U.S. farming industry to rapidly increase its supply of corn and soybeans—the main inputs into biofuels—by developing genetically modified seeds which are resistant to pests and drought. By boosting farm productivity, biotechnology plays a crucial role in ensuring that biofuel production targets are met.

Examples

Biotechnology has applications in four major industrial areas, including health care (medical), crop production and agriculture, non food (industrial) uses of crops and other products (e.g. biodegradable plastics, vegetable oil, biofuels), and environmental uses.

For example, one application of biotechnology is the directed use of organisms for the manufacture of organic products (examples include beer and milk products). Another example is using naturally present bacteria by the mining industry in bioleaching. Biotechnology is also used to recycle, treat waste, clean up sites contaminated by industrial activities (bioremediation), and also to produce biological weapons.

A rose plant that began as cells grown in a tissue culture

A series of derived terms have been coined to identify several branches of biotechnology; for example:

- Bioinformatics is an interdisciplinary field which addresses biological problems using computational techniques, and makes the rapid organization as well as analysis of biological data possible. The field may also be referred to as *computational biology*, and can be defined as, "conceptualizing biology in terms of molecules and then applying informatics techniques to understand and organize the information associated with these molecules, on a large scale." Bioinformatics plays a key role in various areas, such as functional genomics, structural genomics, and proteomics, and forms a key component in the biotechnology and pharmaceutical sector.

- Blue biotechnology is a term that has been used to describe the marine and aquatic applications of biotechnology, but its use is relatively rare.

- Green biotechnology is biotechnology applied to agricultural processes. An example would be the selection and domestication of plants via micropropagation. Another example is the designing of transgenic plants to grow under specific environments in the presence (or absence) of chemicals. One hope is that green biotechnology might produce more environmentally friendly solutions than traditional industrial agriculture. An example of this is the engineering of a plant to express a pesticide, thereby ending the need of external application of pesticides. An example of this would be Bt corn. Whether or not green biotechnology products such as this are ultimately more environmentally friendly is a topic of considerable debate.

- Red biotechnology is applied to medical processes. Some examples are the designing of organisms to produce antibiotics, and the engineering of genetic cures through genetic manipulation.

- White biotechnology, also known as industrial biotechnology, is biotechnology applied to industrial processes. An example is the designing of an organism to produce a useful chemical. Another example is the using of enzymes as industrial catalysts to either produce valuable chemicals or destroy hazardous/polluting chemicals. White biotechnology tends to consume less in resources than traditional processes used to produce industrial goods.

The investment and economic output of all of these types of applied biotechnologies is termed as "bioeconomy".

Medicine

In medicine, modern biotechnology finds applications in areas such as pharmaceutical drug discovery and production, pharmacogenomics, and genetic testing (or genetic screening).

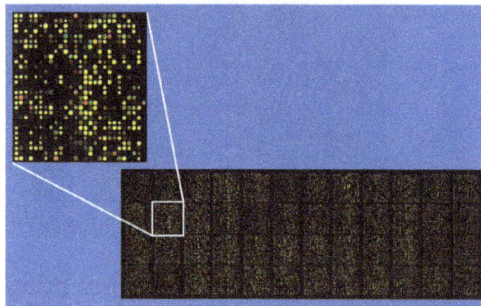

DNA microarray chip – some can do as many as a million blood tests at once

Pharmacogenomics (a combination of pharmacology and genomics) is the technology that analyses how genetic makeup affects an individual's response to drugs. It deals with the influence of genetic variation on drug response in patients by correlating gene expression or single-nucleotide polymorphisms with a drug's efficacy or toxicity. By doing so, pharmacogenomics aims to develop rational means to optimize drug therapy, with respect to the patients' genotype, to ensure maximum efficacy with minimal adverse effects. Such approaches promise the advent of "personalized medicine"; in which drugs and drug combinations are optimized for each individual's unique genetic makeup.

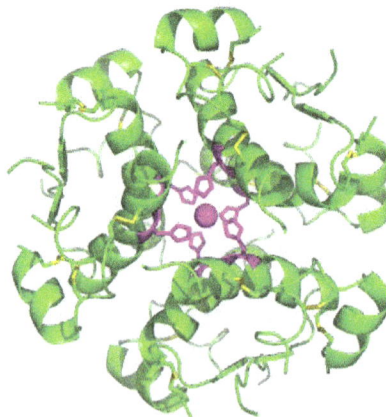

Computer-generated image of insulin hexamers highlighting the threefold symmetry, the zinc ions holding it together, and the histidine residues involved in zinc binding.

Biotechnology has contributed to the discovery and manufacturing of traditional small molecule pharmaceutical drugs as well as drugs that are the product of biotechnology – biopharmaceutics. Modern biotechnology can be used to manufacture existing medicines relatively easily and cheaply. The first genetically engineered products were medicines designed to treat human diseases. To cite one example, in 1978 Genentech developed synthetic humanized insulin by joining its gene with a plasmid vector inserted into the bacterium *Escherichia coli*. Insulin, widely used for the treatment of diabetes, was previously extracted from the pancreas of abattoir animals (cattle and/or pigs). The resulting genetically engineered bacterium enabled the production of vast quantities of synthetic human insulin at relatively low cost. Biotechnology has also enabled emerging therapeutics like gene therapy. The application of biotechnology to basic science (for example through the Human Genome Project) has also dramatically improved our understanding of biology and as our scientific knowledge of normal and disease biology has increased, our ability to develop new medicines to treat previously untreatable diseases has increased as well.

Genetic testing allows the genetic diagnosis of vulnerabilities to inherited diseases, and can also be used to determine a child's parentage (genetic mother and father) or in general a person's ancestry. In addition to studying chromosomes to the level of individual genes, genetic testing in a broader sense includes biochemical tests for the possible presence of genetic diseases, or mutant forms of genes associated with increased risk of developing genetic disorders. Genetic testing identifies changes in chromosomes, genes, or proteins. Most of the time, testing is used to find changes that are associated with inherited disorders. The results of a genetic test can confirm or rule out a suspected genetic condition or help determine a person's chance of developing or passing on a genetic disorder. As of 2011 several hundred genetic tests were in use. Since genetic testing may open up ethical or psychological problems, genetic testing is often accompanied by genetic counseling.

Agriculture

Genetically modified crops ("GM crops", or "biotech crops") are plants used in agriculture, the DNA of which has been modified with genetic engineering techniques. In most cases the aim is to introduce a new trait to the plant which does not occur naturally in the species.

Examples in food crops include resistance to certain pests, diseases, stressful environmental conditions, resistance to chemical treatments (e.g. resistance to a herbicide), reduction of spoilage, or improving the nutrient profile of the crop. Examples in non-food crops include production of pharmaceutical agents, biofuels, and other industrially useful goods, as well as for bioremediation.

Farmers have widely adopted GM technology. Between 1996 and 2011, the total surface area of land cultivated with GM crops had increased by a factor of 94, from 17,000 square kilometers (4,200,000 acres) to 1,600,000 km² (395 million acres). 10% of the world's crop lands were planted with GM crops in 2010. As of 2011, 11 different transgenic crops were grown commercially on 395 million acres (160 million hectares) in 29 countries such as the USA, Brazil, Argentina, India, Canada, China, Paraguay, Pakistan, South Africa, Uruguay, Bolivia, Australia, Philippines, Myanmar, Burkina Faso, Mexico and Spain.

Genetically modified foods are foods produced from organisms that have had specific changes introduced into their DNA with the methods of genetic engineering. These techniques have allowed for the introduction of new crop traits as well as a far greater control over a food's genetic structure than previously afforded by methods such as selective breeding and mutation breeding. Commercial sale of genetically modified foods began in 1994, when Calgene first marketed its Flavr Savr delayed ripening tomato. To date most genetic modification of foods have primarily focused on cash crops in high demand by farmers such as soybean, corn, canola, and cotton seed oil. These have been engineered for resistance to pathogens and herbicides and better nutrient profiles. GM livestock have also been experimentally developed, although as of November 2013 none are currently on the market.

There is a scientific consensus that currently available food derived from GM crops poses no greater risk to human health than conventional food, but that each GM food needs to be tested on a case-by-case basis before introduction. Nonetheless, members of the public are much less likely than scientists to perceive GM foods as safe. The legal and regulatory status of GM foods varies by country, with some nations banning or restricting them, and others permitting them with widely differing degrees of regulation.

GM crops also provide a number of ecological benefits, if not used in excess. However, opponents have objected to GM crops per se on several grounds, including environmental concerns, whether food produced from GM crops is safe, whether GM crops are needed to address the world's food needs, and economic concerns raised by the fact these organisms are subject to intellectual property law.

Industrial

Industrial biotechnology (known mainly in Europe as white biotechnology) is the application of biotechnology for industrial purposes, including industrial fermentation. It includes the practice of using cells such as micro-organisms, or components of cells like enzymes, to generate industrially useful products in sectors such as chemicals, food and feed, detergents, paper and pulp, textiles and biofuels. In doing so, biotechnology uses renewable raw materials and may contribute to lowering greenhouse gas emissions and moving away from a petrochemical-based economy.

Environmental

The environment can be affected by biotechnologies, both positively and adversely. Vallero and others have argued that the difference between beneficial biotechnology (e.g. bioremediation to clean up an oil spill or hazard chemical leak) versus the adverse effects stemming from biotechnological enterprises (e.g. flow of genetic material from transgenic organisms into wild strains) can be seen as applications and implications, respectively. Cleaning up environmental wastes is an example of an application of environmental biotechnology; whereas loss of biodiversity or loss of containment of a harmful microbe are examples of environmental implications of biotechnology.

Regulation

The regulation of genetic engineering concerns approaches taken by governments to assess and manage the risks associated with the use of genetic engineering technology, and the development and release of genetically modified organisms (GMO), including genetically modified crops and genetically modified fish. There are differences in the regulation of GMOs between countries, with some of the most marked differences occurring between the USA and Europe. Regulation varies in a given country depending on the intended use of the products of the genetic engineering. For example, a crop not intended for food use is generally not reviewed by authorities responsible for food safety. The European Union differentiates between approval for cultivation within the EU and approval for import and processing. While only a few GMOs have been approved for cultivation in the EU a number of GMOs have been approved for import and processing. The cultivation of GMOs has triggered a debate about coexistence of GM and non GM crops. Depending on the coexistence regulations incentives for cultivation of GM crops differ.

Learning

In 1988, after prompting from the United States Congress, the National Institute of General Medical Sciences (National Institutes of Health) (NIGMS) instituted a funding mechanism for biotechnology training. Universities nationwide compete for these funds to establish Biotechnology Training Programs (BTPs). Each successful application is generally funded for five years then must be competitively renewed. Graduate students in turn compete for acceptance into a BTP; if accepted, then stipend, tuition and health insurance support is provided for two or three years during the course of their Ph.D. thesis work. Nineteen institutions offer NIGMS supported BTPs. Biotechnology training is also offered at the undergraduate level and in community colleges.

References

- Thieman, W.J.; Palladino, M.A. (2008). Introduction to Biotechnology. Pearson/Benjamin Cummings. ISBN 0-321-49145-9.

- Springham, D.; Springham, G.; Moses, V.; Cape, R.E. (24 August 1999). Biotechnology: The Science and the Business. CRC Press. p. 1. ISBN 978-90-5702-407-8.

- Diaz E (editor). (2008). Microbial Biodegradation: Genomics and Molecular Biology (1st ed.). Caister Academic Press. ISBN 1-904455-17-4.

- "A decade of EU-funded GMO research (2001–2010)" (PDF). Directorate-General for Research and Innovation. Biotechnologies, Agriculture, Food. European Commission, European Union. 2010. doi:10.2777/97784. ISBN 978-92-79-16344-9. Retrieved February 8, 2016.

- John P. (2005) [1911]. Origin and History of Beer and Brewing: From Prehistoric Times to the Beginning of Brewing Science and Technology. Cleveland, Ohio: BeerBooks. p. 34. ISBN 978-0-9662084-1-2. OCLC 71834130.

- Final Report of the PABE research project (December 2001). "Public Perceptions of Agricultural Biotechnologies in Europe". Commission of European Communities. Retrieved February 24, 2016.

- Bashshur, Ramona (February 2013). "FDA and Regulation of GMOs". American Bar Association. Retrieved February 24, 2016.

- Lynch, Diahanna; Vogel, David (April 5, 2001). "The Regulation of GMOs in Europe and the United States: A Case-Study of Contemporary European Regulatory Politics". Council on Foreign Relations. Retrieved February 24, 2016.

- Carrington, Damien (19 January 2012) GM microbe breakthrough paves way for large-scale seaweed farming for biofuels The Guardian. Retrieved 12 March 2012

- James, C (2011). "ISAAA Brief 43, Global Status of Commercialized Biotech/GM Crops: 2011". ISAAA Briefs. Ithaca, New York: International Service for the Acquisition of Agri-biotech Applications (ISAAA). Retrieved 2012-06-02.

- "Incorporating Biotechnology into the Classroom What is Biotechnology?", from the curricula of the 'Incorporating Biotechnology into the High School Classroom through Arizona State University's BioREACH program', accessed on October 16, 2012). Public.asu.edu. Retrieved on 2013-03-20.

- Paarlburg, Robert Drought Tolerant GMO Maize in Africa, Anticipating Regulatory Hurdles International Life Sciences Institute, January 2011. Retrieved 25 April 2011

- James, Clive (1996). "Global Review of the Field Testing and Commercialization of Transgenic Plants: 1986 to 1995" (PDF). The International Service for the Acquisition of Agri-biotech Applications. Retrieved 17 July 2010.

2

Concepts of Biotechnology

Biobased economy, green revolution, biomimetic or tissue culture are important concepts of biotechnology dealt within this chapter. The following chapter unfolds the crucial aspects in a critical yet systematic manner and elucidates the crucial theories and principles of biotechnology. The major concepts of biotechnology are discussed in this chapter.

Biobased Economy

Biobased economy, bioeconomy or biotechonomy refers to all economic activity derived from scientific and research activity focused on biotechnology. In other words, understanding mechanisms and processes at the genetic and molecular levels and applying this understanding to creating or improving industrial processes.

The term is widely used by regional development agencies, international organizations, biotechnology companies. It is closely linked to the evolution of the biotechnology industry. The ability to study, understand and manipulate genetic material has been possible due to scientific breakthroughs and technological progress.

The evolution of the biotechnology industry and its application to agriculture, health, chemical or energy industries is a classic example of bioeconomic activity.

History

The term was first defined by Juan Enríquez and Rodrigo Martinez, Life Sciences Chief Strategist at IDEO at the Genomics Seminar in the 1997 AAAS meeting. An excerpt of this paper was published in *Science*."

Enríquez and Martinez' 2002 Harvard Business School working paper, "Biotechonomy 1.0: A Rough Map of Biodata Flow", showed the global flow of genetic material into and out of the three largest public genetic databases: GenBank, EMBL and DDBJ. The authors then hypothesized about the economic impact that such data flows might have on patent creation, evolution of biotech startups and licensing fees. An adaptation of this paper was published in *Wired* magazine in 2003.

Since about 2005, the creation of a biobased economy has been a significant issue in the Netherlands. Pilot plants have been started i.e. in Lelystad (Zeafuels), and a centralised organisation exists (Interdepartementaal programma biobased economy), with supporting research (Food & Biobased Research) being conducted.

In 2012 president Barack Obama of the USA announced intentions to encourage biological manufacturing methods, with a National Bioeconomy Blueprint. In the same year in Belgium the Bio Base Europe Pilot Plant was set up.

In Practice

The biobased economy uses first-generation biomass (crops), second-generation biomass (crop refuge), and third-generation biomass (seaweed, algae). Several methods of processing are then used to gather the most out of the biomass. This includes techniques such as

- Anaerobic digestion

- Pyrolysis

- Torrefaction

- Fermentation

- Biorefinery

Anaerobic digestion is generally used to produce ethanol, pyrolysis is used to produce pyrolysis-oil (which is solidified biogas), and torrefaction is used to create biomass-coal. Biomass-coal and biogas is then burnt for energy production, ethanol can be used as a (vehicle)-fuel, as well as for other purposes, such as skincare products.

Getting the Most Out of the Biomass

For economic reasons, the processing of the biomass is done according to a specific pattern. This pattern, as well as the quantities, depends on the types of biomass used. The whole of finding the most suitable pattern is known as biorefining. A general list shows the products with high added value and lowest volume of biomass to the products with the lowest added value and highest volume of biomass:

- fine chemicals/medicines

- food

- chemicals/bioplastics

- transport fuels

- electricity and heat

Some research is being conducted as well in order to improve the manufacturing processes. For example, to make plastics, paint, medicines, antifreeze out of syngas, a new catalyst has been invented by Krijn de Jong.

Comparison of Fossil Fuel and Biobased Economy

- With a fossil fuel economy substances as gasoline, fuel oil, diesel, naphta, kerosine, LPG, and other are converted to: energy, chemical products, food, materials

- With a biobased economy substances as (syn)gas, sugars, oil, fibres and other are converted to energy, chemical products, (animal) food, biomaterials

Green Revolution

After the Second World War, increased deployment of technologies including pesticides, herbicides, and fertilizers as well as new breeds of high yield crops greatly increased global food production .

The Green Revolution refers to a set of research and development of technology transfer initiatives occurring between the 1930s and the late 1960s (with prequels in the work of the agrarian geneticist Nazareno Strampelli in the 1920s and 1930s), that increased agricultural production worldwide, particularly in the developing world, beginning most markedly in the late 1960s. The initiatives resulted in the adoption of new technologies, including:

...new, high-yielding varieties (HYVs) of cereals, especially dwarf wheats and rices, in association with chemical fertilizers and agro-chemicals, and with controlled water-supply (usually involving irrigation) and new methods of cultivation, including mechanization. All of these together were seen as a 'package of practices' to supersede 'traditional' technology and to be adopted as a whole.

The initiatives, led by Norman Borlaug, the "Father of the Green Revolution," who received the Nobel Peace Prize in 1970, credited with saving over a billion people from starvation, involved the development of high-yielding varieties of cereal grains, expansion of irrigation infrastructure, modernization of management techniques, distribution of hybridized seeds, synthetic fertilizers, and pesticides to farmers.

The term "Green Revolution" was first used in 1968 by former US Agency for International Development (USAID) director William Gaud, who noted the spread of the new technologies: "These and other developments in the field of agriculture contain the makings of a new revolution. It is not a violent Red Revolution like that of the Soviets, nor is it a White Revolution like that of the Shah of Iran. I call it the Green Revolution."

History

Green Revolution in Mexico

It has been argued that during the twentieth century two 'revolutions' transformed rural Mexico: the Mexican Revolution (1910-1920) and the Green Revolution (1950-1970). With the support of the Mexican government, the U.S. government, the United Nations, the Food and Agriculture Organization (FAO), and the Rockefeller Foundation, Mexico made a concerted effort to transform agricultural productivity, particularly with irrigated rather than dry-land cultivation in its northwest, to solve its problem of lack of food self-sufficiency. In the center and south of Mexico, where large-scale production faced challenges, agricultural production languished. Increased production meant food self-sufficiency in Mexico to feed its growing and urbanizing population, with the number of calories consumed per Mexican increasing.

Mexico was not merely the recipient of Green Revolution knowledge and technology, but was an active participant with financial support from the government for agriculture as well as Mexican agronomists (*agrónomos*). Although the Mexican Revolution had broken the back of the hacienda system and land reform in Mexico had by 1940 distributed a large expanse of land in central and southern Mexico, agricultural productivity had fallen. During the administration of Manuel Avila Camacho (1940–46), the government put resources into developing new breeds of plants and partnered with the Rockefeller Foundation. In 1943, the Mexican government founded the International Maize and Wheat Improvement Center (CIMMYT), which became a base for international agricultural research.

Agriculture in Mexico had been a sociopolitical issue, a key factor in some regions' participation in the Mexican Revolution. It was also a technical issue, which the development of a cohort trained agronomists, who were to advise peasants how to increase productivity. In the post-World War II era, the government sought development in agriculture that bettered technological aspects of agriculture in regions that were not dominated by small-scale peasant cultivators. This drive for transforming agriculture would have the benefit of keeping Mexico self-sufficient in food and in the political sphere with the Cold War, potentially stem unrest and the appeal of Communism. Technical aid can be seen as also serving political ends in the international sphere. In Mexico, it also served political ends, separating peasant agriculture based on the ejido and considered one of the victories of the Mexican Revolution, from agribusiness that requires large-scale land ownership, irrigation, specialized seeds, fertilizers, and pesticides, machinery, and a low-wage paid labor force.

The government created the Mexican Agricultural Program (MAP) to be the lead organization in raising productivity. One of their successes was wheat production, with varieties the agency's scientists helped create dominating wheat production as early as 1951 (70%), 1965 (80%), and 1968 (90%). Mexico became the showcase for extending the Green Revolution to other areas of Latin America and beyond, into Africa and Asia. New breeds of maize, beans, along with wheat produced bumper crops with proper inputs (such as fertilizer and pesticides) and careful cultivation. Many

Mexican farmers who had been dubious about the scientists or hostile to them (often a mutual relationship of discord) came to see the scientific approach to agriculture worth adopting.

Green Revolution in Rice: IR8 and the Philippines

In 1960, the Government of the Republic of the Philippines with the Ford Foundation and the Rockefeller Foundation established IRRI (International Rice Research Institute). A rice crossing between Dee-Geo-woo-gen and Peta was done at IRRI in 1962. In 1966, one of the breeding lines became a new cultivar, IR8. IR8 required the use of fertilizers and pesticides, but produced substantially higher yields than the traditional cultivars. Annual rice production in the Philippines increased from 3.7 to 7.7 million tons in two decades. The switch to IR8 rice made the Philippines a rice exporter for the first time in the 20th century.

Green Revolution's Start in India

In 1961, India was on the brink of mass famine. Norman Borlaug was invited to India by the adviser to the Indian minister of agriculture C. Subramaniam. Despite bureaucratic hurdles imposed by India's grain monopolies, the Ford Foundation and Indian government collaborated to import wheat seed from the International Maize and Wheat Improvement Center (CIMMYT). Punjab was selected by the Indian government to be the first site to try the new crops because of its reliable water supply and a history of agricultural success. India began its own Green Revolution program of plant breeding, irrigation development, and financing of agrochemicals.

India soon adopted IR8 – a semi-dwarf rice variety developed by the International Rice Research Institute (IRRI) that could produce more grains of rice per plant when grown with certain fertilizers and irrigation. In 1968, Indian agronomist S.K. De Datta published his findings that IR8 rice yielded about 5 tons per hectare with no fertilizer, and almost 10 tons per hectare under optimal conditions. This was 10 times the yield of traditional rice. IR8 was a success throughout Asia, and dubbed the "Miracle Rice". IR8 was also developed into Semi-dwarf IR36.

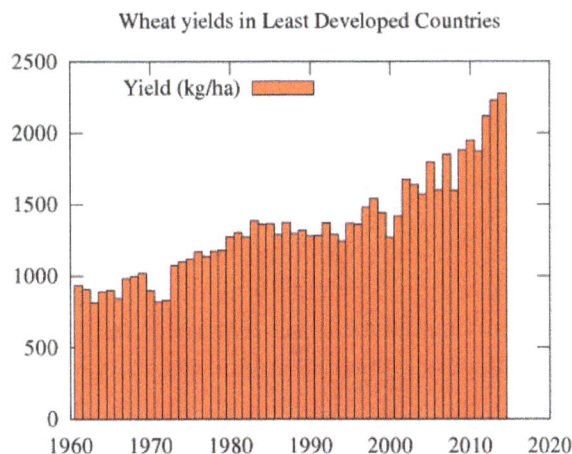

Wheat yields in least developed countries since 1961, in kilograms per hectare.

In the 1960s, rice yields in India were about two tons per hectare; by the mid-1990s, they had risen to six tons per hectare. In the 1970s, rice cost about $550 a ton; in 2001, it cost under $200 a ton.

India became one of the world's most successful rice producers, and is now a major rice exporter, shipping nearly 4.5 million tons in 2006.

Consultative Group on International Agricultural Research - CGIAR

In 1970, foundation officials proposed a worldwide network of agricultural research centers under a permanent secretariat. This was further supported and developed by the World Bank; on 19 May 1971, the Consultative Group on International Agricultural Research(CGIAR) was established, co-sponsored by the FAO, IFAD and UNDP. CGIAR, has added many research centers throughout the world.

CGIAR has responded, at least in part, to criticisms of Green Revolution methodologies. This began in the 1980s, and mainly was a result of pressure from donor organizations. Methods like Agroecosystem Analysis and Farming System Research have been adopted to gain a more holistic view of agriculture.

Brazil's Agricultural Revolution

Brazil's vast inland cerrado region was regarded as unfit for farming before the 1960s because the soil was too acidic and poor in nutrients, according to Norman Borlaug. However, from the 1960s, vast quantities of lime (pulverised chalk or limestone) were poured on the soil to reduce acidity. The effort went on and in the late 1990s between 14 million and 16 million tonnes of lime were being spread on Brazilian fields each year. The quantity rose to 25 million tonnes in 2003 and 2004, equalling around five tonnes of lime per hectare. As a result, Brazil has become the world's second biggest soybean exporter and, thanks to the boom in animal feed production, Brazil is now the biggest exporter of beef and poultry in the world.

Problems in Africa

There have been numerous attempts to introduce the successful concepts from the Mexican and Indian projects into Africa. These programs have generally been less successful. Reasons cited include widespread corruption, insecurity, a lack of infrastructure, and a general lack of will on the part of the governments. Yet environmental factors, such as the availability of water for irrigation, the high diversity in slope and soil types in one given area are also reasons why the Green Revolution is not so successful in Africa.

A recent program in western Africa is attempting to introduce a new high-yielding 'family' of rice varieties known as "New Rice for Africa" (NERICA). NERICA varieties yield about 30% more rice under normal conditions, and can double yields with small amounts of fertilizer and very basic irrigation. However, the program has been beset by problems getting the rice into the hands of farmers, and to date the only success has been in Guinea, where it currently accounts for 16% of rice cultivation.

After a famine in 2001 and years of chronic hunger and poverty, in 2005 the small African country of Malawi launched the "Agricultural Input Subsidy Program" by which vouchers are given to smallholder farmers to buy subsidized nitrogen fertilizer and maize seeds. Within its first year, the program was reported to have had extreme success, producing the largest maize harvest of the country's history, enough to feed the country with tons of maize left over. The program has advanced yearly ever since. Various sources claim that the program has been an unusual success, hailing it as a "miracle".

Agricultural Production and Food Security

Technologies

New varieties of wheat and other grains were instrumental to the green revolution.

The Green Revolution spread technologies that already existed, but had not been widely implemented outside industrialized nations. These technologies included modern irrigation projects, pesticides, synthetic nitrogen fertilizer and improved crop varieties developed through the conventional, science-based methods available at the time.

The novel technological development of the Green Revolution was the production of novel wheat cultivars. Agronomists bred cultivars of maize, wheat, and rice that are generally referred to as HYVs or "high-yielding varieties". HYVs have higher nitrogen-absorbing potential than other varieties. Since cereals that absorbed extra nitrogen would typically lodge, or fall over before harvest, semi-dwarfing genes were bred into their genomes. A Japanese dwarf wheat cultivar (Norin 10 wheat), which was sent to Washington, D.C. by Cecil Salmon, was instrumental in developing Green Revolution wheat cultivars. IR8, the first widely implemented HYV rice to be developed by IRRI, was created through a cross between an Indonesian variety named "Peta" and a Chinese variety named "Dee-geo-woo-gen."

With advances in molecular genetics, the mutant genes responsible for *Arabidopsis thaliana* genes (GA 20-oxidase, *ga1*, *ga1-3*), wheat reduced-height genes (*Rht*) and a rice semidwarf gene (*sd1*) were cloned. These were identified as gibberellin biosynthesis genes or cellular signaling component genes. Stem growth in the mutant background is significantly reduced leading to the dwarf phenotype. Photosynthetic investment in the stem is reduced dramatically as the shorter plants are inherently more stable mechanically. Assimilates become redirected to grain production, amplifying in particular the effect of chemical fertilizers on commercial yield.

HYVs significantly outperform traditional varieties in the presence of adequate irrigation, pesticides, and fertilizers. In the absence of these inputs, traditional varieties may outperform HYVs.

Therefore, several authors have challenged the apparent superiority of HYVs not only compared to the traditional varieties alone, but by contrasting the monocultural system associated with HYVs with the polycultural system associated with traditional ones.

Production Increases

Cereal production more than doubled in developing nations between the years 1961–1985. Yields of rice, maize, and wheat increased steadily during that period. The production increases can be attributed roughly equally to irrigation, fertilizer, and seed development, at least in the case of Asian rice.

While agricultural output increased as a result of the Green Revolution, the energy input to produce a crop has increased faster, so that the ratio of crops produced to energy input has decreased over time. Green Revolution techniques also heavily rely on chemical fertilizers, pesticides and herbicides and rely on machines, which as of 2014 rely on or are derived from crude oil, making agriculture increasingly reliant on crude oil extraction. Proponents of the Peak Oil theory fear that a future decline in oil and gas production would lead to a decline in food production or even a Malthusian catastrophe.

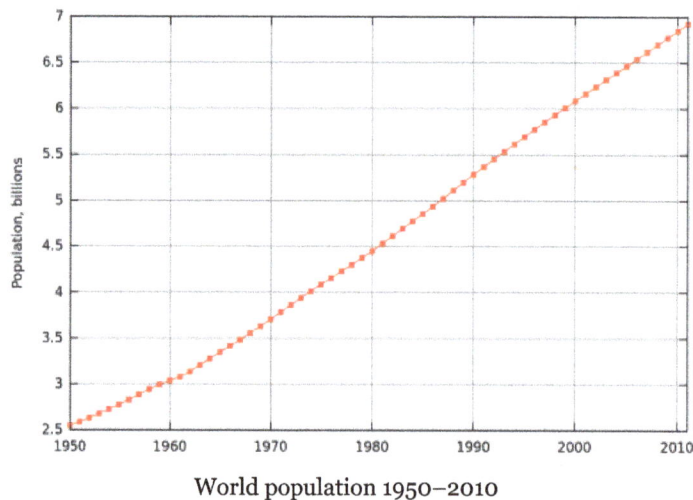

World population 1950–2010

Effects on Food Security

The effects of the Green Revolution on global food security are difficult to assess because of the complexities involved in food systems.

The world population has grown by about four billion since the beginning of the Green Revolution and many believe that, without the Revolution, there would have been greater famine and malnutrition. India saw annual wheat production rise from 10 million tons in the 1960s to 73 million in 2006. The average person in the developing world consumes roughly 25% more calories per day now than before the Green Revolution. Between 1950 and 1984, as the Green Revolution transformed agriculture around the globe, world grain production increased by about 160%.

The production increases fostered by the Green Revolution are often credited with having helped to avoid widespread famine, and for feeding billions of people.

There are also claims that the Green Revolution has decreased food security for a large number of people. One claim involves the shift of subsistence-oriented cropland to cropland oriented towards production of grain for export or animal feed. For example, the Green Revolution replaced much of the land used for pulses that fed Indian peasants for wheat, which did not make up a large portion of the peasant diet.

Criticism

3rd World Economic Sovereignty

A main criticism of the effects of the green revolution is the cost for many small farmers using HYV seeds, with their associated demands of increased irrigation systems and pesticides. A case study is found in India, where farmers are planting cotton seeds capable of producing Bt toxin. A criticism regarding the green revolution are the effects regarding the widespread commercialization and market share of organisations, particularly of the phasing out of seed saving practices in favor of purchasing of seeds, and concerns regarding the financial affordability of the adoption of patented crops amongst farmers, particularly of those in the developing world. This can allow larger farms, even foreign owned farming operations, to buy up local smallhold farms.

Vandana Shiva notes that this is the "second Green Revolution". The first Green Revolution, she notes, was mostly publicly-funded (by the Indian Government). This new Green Revolution, she says, is driven by private [and foreign] interest - notably MNCs like Monsanto. Ultimately, this is leading to foreign ownership over most of India's farmland.

Food Security

Malthusian Criticism

Some criticisms generally involve some variation of the Malthusian principle of population. Such concerns often revolve around the idea that the Green Revolution is unsustainable, and argue that humanity is now in a state of overpopulation or overshoot with regards to the sustainable carrying capacity and ecological demands on the Earth.

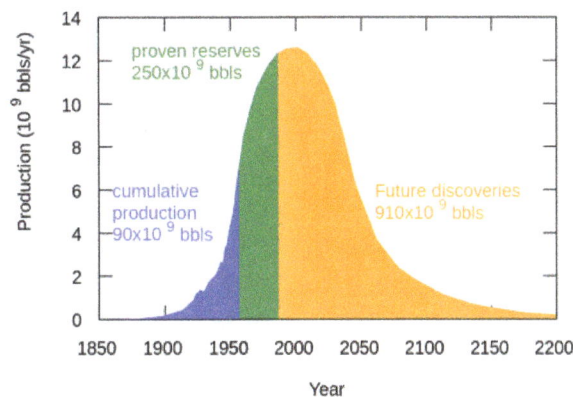

M. King Hubbert's prediction of world petroleum production rates. Modern agriculture is largely reliant on petroleum energy.

Although 36 million people die each year as a direct or indirect result of hunger and poor nutrition,

Malthus's more extreme predictions have frequently failed to materialize. In 1798 Thomas Malthus made his prediction of impending famine. The world's population had doubled by 1923 and doubled again by 1973 without fulfilling Malthus's prediction. Malthusian Paul R. Ehrlich, in his 1968 book *The Population Bomb*, said that "India couldn't possibly feed two hundred million more people by 1980" and "Hundreds of millions of people will starve to death in spite of any crash programs." Ehrlich's warnings failed to materialize when India became self-sustaining in cereal production in 1974 (six years later) as a result of the introduction of Norman Borlaug's dwarf wheat varieties.

Since supplies of oil and gas are essential to modern agriculture techniques, a fall in global oil supplies could cause spiking food prices in the coming decades.

Famine

To some modern Western sociologists and writers, increasing food production is not synonymous with increasing food security, and is only part of a larger equation. For example, Harvard professor Amartya Sen claimed large historic famines were not caused by decreases in food supply, but by socioeconomic dynamics and a failure of public action. However, economist Peter Bowbrick disputes Sen's theory, arguing that Sen relies on inconsistent arguments and contradicts available information, including sources that Sen himself cited. Bowbrick further argues that Sen's views coincide with that of the Bengal government at the time of the Bengal famine of 1943, and the policies Sen advocates failed to relieve the famine.

Quality of Diet

Some have challenged the value of the increased food production of Green Revolution agriculture. Miguel A. Altieri, (a pioneer of agroecology and peasant-advocate), writes that the comparison between traditional systems of agriculture and Green Revolution agriculture has been unfair, because Green Revolution agriculture produces monocultures of cereal grains, while traditional agriculture usually incorporates polycultures.

These monoculture crops are often used for export, feed for animals, or conversion into biofuel. According to Emile Frison of Bioversity International, the Green Revolution has also led to a change in dietary habits, as fewer people are affected by hunger and die from starvation, but many are affected by malnutrition such as iron or vitamin-A deficiencies. Frison further asserts that almost 60% of yearly deaths of children under age five in developing countries are related to malnutrition.

High-yield rice (HYR), introduced since 1964 to poverty-ridden Asian countries, such as the Philippines, was found to have inferior flavor and be more glutinous and less savory than their native varieties. This caused its price to be lower than the average market value.

In the Philippines the introduction of heavy pesticides to rice production, in the early part of the Green Revolution, poisoned and killed off fish and weedy green vegetables that traditionally coexisted in rice paddies. These were nutritious food sources for many poor Filipino farmers prior to the introduction of pesticides, further impacting the diets of locals.

Political Impact

A major critic of the Green Revolution, U.S. investigative journalist Mark Dowie, writes:

The primary objective of the program was geopolitical: to provide food for the populace in undeveloped countries and so bring social stability and weaken the fomenting of communist insurgency.

Citing internal Foundation documents, Dowie states that the Ford Foundation had a greater concern than Rockefeller in this area.

There is significant evidence that the Green Revolution weakened socialist movements in many nations. In countries such as India, Mexico, and the Philippines, *technological solutions* were sought as an alternative to expanding *agrarian reform* initiatives, the latter of which were often linked to socialist politics.

Socioeconomic Impacts

The transition from traditional agriculture, in which inputs were generated on-farm, to Green Revolution agriculture, which required the purchase of inputs, led to the widespread establishment of rural credit institutions. Smaller farmers often went into debt, which in many cases results in a loss of their farmland. The increased level of mechanization on larger farms made possible by the Green Revolution removed a large source of employment from the rural economy. Because wealthier farmers had better access to credit and land, the Green Revolution increased class disparities, with the rich–poor gap widening as a result. Because some regions were able to adopt Green Revolution agriculture more readily than others (for political or geographical reasons), interregional economic disparities increased as well. Many small farmers are hurt by the dropping prices resulting from increased production overall. However, large-scale farming companies only account for less than 10% of the total farming capacity. This is a criticism held by many small producers in the food sovereignty movement.

The new economic difficulties of small holder farmers and landless farm workers led to increased rural-urban migration. The increase in food production led to a cheaper food for urban dwellers, and the increase in urban population increased the potential for industrialization.

Globalization

In the most basic sense, the Green Revolution was a product of globalization as evidenced in the creation of international agricultural research centers that shared information, and with transnational funding from groups like the Rockefeller Foundation, Ford Foundation, and United States Agency for International Development (USAID).

Environmental Impact

Increased use of irrigation played a major role in the green revolution.

Biodiversity

The spread of Green Revolution agriculture affected both agricultural biodiversity (or agrodiversity) and wild biodiversity. There is little disagreement that the Green Revolution acted to reduce agricultural biodiversity, as it relied on just a few high-yield varieties of each crop.

This has led to concerns about the susceptibility of a food supply to pathogens that cannot be controlled by agrochemicals, as well as the permanent loss of many valuable genetic traits bred into traditional varieties over thousands of years. To address these concerns, massive seed banks such as Consultative Group on International Agricultural Research's (CGIAR) International Plant Genetic Resources Institute (now Bioversity International) have been established.

There are varying opinions about the effect of the Green Revolution on wild biodiversity. One hypothesis speculates that by increasing production per unit of land area, agriculture will not need to expand into new, uncultivated areas to feed a growing human population. However, land degradation and soil nutrients depletion have forced farmers to clear up formerly forested areas in order to keep up with production. A counter-hypothesis speculates that biodiversity was sacrificed because traditional systems of agriculture that were displaced sometimes incorporated practices to preserve wild biodiversity, and because the Green Revolution expanded agricultural development into new areas where it was once unprofitable or too arid. For example, the development of wheat varieties tolerant to acid soil conditions with high aluminium content, permitted the introduction of agriculture in sensitive Brazilian ecosystems as Cerrado semi-humid tropical savanna and Amazon rainforest in the geoeconomic macroregions of Centro-Sul and Amazônia. Before the Green Revolution, other Brazilian ecosystems were also significantly damaged by human activity, such as the once 1st or 2nd main contributor to Brazilian megadiversity Atlantic Rainforest (above 85% of deforestation in the 1980s, about 95% after the 2010s) and the important xeric shrublands called Caatinga mainly in the Northeastern Brazil (about 40% in the 1980s, about 50% after the 2010s — deforestation of the Caatinga biome is generally associated with greater risks of desertification).

Nevertheless, the world community has clearly acknowledged the negative aspects of agricultural expansion as the 1992 Rio Treaty, signed by 189 nations, has generated numerous national Biodiversity Action Plans which assign significant biodiversity loss to agriculture's expansion into new domains.

Greenhouse Gas Emissions

According to a study published in 2013 in PNAS, in the absence of the crop germplasm improvement associated with the Green revolution, greenhouse gas emissions would have been 5.2-7.4 Gt higher than observed in 1965–2004.

Dependence on Non-renewable Resources

Most high intensity agricultural production is highly reliant on non-renewable resources. Agricultural machinery and transport, as well as the production of pesticides and nitrates all depend on fossil fuels. Moreover, the essential mineral nutrient phosphorus is often a limiting factor in crop cultivation, while phosphorus mines are rapidly being depleted worldwide. The failure to depart from these non-sustainable agricultural production methods could potentially lead to a large scale collapse of the current system of intensive food production within this century.

Health Impact

The consumption of the pesticides used to kill pests by humans in some cases may be increasing the likelihood of cancer in some of the rural villages using them. Poor farming practices including non-compliance to usage of masks and over-usage of the chemicals compound this situation. In 1989, WHO and UNEP estimated that there were around 1 million human pesticide poisonings annually. Some 20,000 (mostly in developing countries) ended in death, as a result of poor labeling, loose safety standards etc.

Pesticides and Cancer

Long term exposure to pesticides such as organochlorines, creosote, and sulfate have been correlated with higher cancer rates and organochlorines DDT, chlordane, and lindane as tumor promoters in animals. Contradictory epidemiologic studies in humans have linked phenoxy acid herbicides or contaminants in them with soft tissue sarcoma (STS) and malignant lymphoma, organochlorine insecticides with STS, non-Hodgkin's lymphoma (NHL), leukemia, and, less consistently, with cancers of the lung and breast, organophosphorous compounds with NHL and leukemia, and triazine herbicides with ovarian cancer.

Punjab Case

The Indian state of Punjab pioneered green revolution among the other states transforming India into a food-surplus country. The state is witnessing serious consequences of intensive farming using chemicals and pesticide. A comprehensive study conducted by Post Graduate Institute of Medical Education and Research (PGIMER) has underlined the direct relationship between indiscriminate use of these chemicals and increased incidence of cancer in this region. An increase in the number of cancer cases has been reported in several villages including Jhariwala, Koharwala, Puckka, Bhimawali, and Khara.

Environmental activist Vandana Shiva has written extensively about the social, political and economic impacts of the Green Revolution in Punjab. She claims that the Green Revolution's reliance on heavy use of chemical inputs and monocultures has resulted in water scarcity, vulnerability to pests, and incidents of violent conflict and social marginalization.

In 2009, under a Greenpeace Research Laboratories investigation, Dr Reyes Tirado, from the University of Exeter, UK conducted the study in 50 villages in Muktsar, Bathinda and Ludhiana districts revealed chemical, radiation and biological toxicity rampant in Punjab. Twenty percent of the sampled wells showed nitrate levels above the safety limit of 50 mg/l, established by WHO, the study connected it with high use of synthetic nitrogen fertilizers.

Norman Borlaug's Response to Criticism

Borlaug dismissed certain claims of critics, but also cautioned "There are no miracles in agricultural production. Nor is there such a thing as a miracle variety of wheat, rice, or maize which can serve as an elixir to cure all ills of a stagnant, traditional agriculture."

Of environmental lobbyists, he said:

"some of the environmental lobbyists of the Western nations are the salt of the earth, but many of them are elitists. They've never experienced the physical sensation of hunger. They do their lobbying from comfortable office suites in Washington or Brussels...If they lived just one month amid the misery of the developing world, as I have for fifty years, they'd be crying out for tractors and fertilizer and irrigation canals and be outraged that fashionable elitists back home were trying to deny them these things".

However, the charge of "elitism" could also be leveled against supporters of the Green Revolution. As noted in the documentary Profits from Poison, protective gear for farmers is often designed by people who have never experienced tropical climates. For example, plastic ponchos create a sauna effect if worn in high temperature/humidity, unlike the experience of wearing them in an airconditioned office.

The "New" Green Revolution

Although the Green Revolution has been able to improve agricultural output in some regions in the world, there was and is still room for improvement. As a result, many organizations continue to invent new ways to improve the techniques already used in the Green Revolution. Frequently quoted inventions are the System of Rice Intensification, marker-assisted selection, agroecology, and applying existing technologies to agricultural problems of the developing world.

Stem Cell

Stem cells are undifferentiated biological cells that can differentiate into specialized cells and can divide (through mitosis) to produce more stem cells. They are found in multicellular organisms. In mammals, there are two broad types of stem cells: embryonic stem cells, which are isolated from the inner cell mass of blastocysts, and adult stem cells, which are found in various tissues. In adult organisms, stem cells and progenitor cells act as a repair system for the body, replenishing adult tissues. In a developing embryo, stem cells can differentiate into all the specialized cells—ectoderm, endoderm and mesoderm—but also maintain the normal turnover of regenerative organs, such as blood, skin, or intestinal tissues.

There are three known accessible sources of autologous adult stem cells in humans:

1. Bone marrow, which requires extraction by *harvesting*, that is, drilling into bone (typically the femur or iliac crest).

2. Adipose tissue (lipid cells), which requires extraction by liposuction.

3. Blood, which requires extraction through apheresis, wherein blood is drawn from the donor (similar to a blood donation), and passed through a machine that extracts the stem cells and returns other portions of the blood to the donor.

Stem cells can also be taken from umbilical cord blood just after birth. Of all stem cell types, autologous harvesting involves the least risk. By definition, autologous cells are obtained from one's own body, just as one may bank his or her own blood for elective surgical procedures.

Adult stem cells are frequently used in various medical therapies (e.g., bone marrow transplantation). Stem cells can now be artificially grown and transformed (differentiated) into specialized cell types with characteristics consistent with cells of various tissues such as muscles or nerves. Embryonic cell lines and autologous embryonic stem cells generated through somatic cell nuclear transfer or dedifferentiation have also been proposed as promising candidates for future therapies. Research into stem cells grew out of findings by Ernest A. McCulloch and James E. Till at the University of Toronto in the 1960s.

Properties

The classical definition of a stem cell requires that it possess two properties:

- *Self-renewal*: the ability to go through numerous cycles of cell division while maintaining the undifferentiated state.

- *Potency*: the capacity to differentiate into specialized cell types. In the strictest sense, this requires stem cells to be either totipotent or pluripotent—to be able to give rise to any mature cell type, although multipotent or unipotent progenitor cells are sometimes referred to as stem cells. Apart from this it is said that stem cell function is regulated in a feed back mechanism.

Self-renewal

Two mechanisms exist to ensure that a stem cell population is maintained:

1. Obligatory asymmetric replication: a stem cell divides into one mother cell that is identical to the original stem cell, and another daughter cell that is differentiated.

2. Stochastic differentiation: when one stem cell develops into two differentiated daughter cells, another stem cell undergoes mitosis and produces two stem cells identical to the original.

Potency Definition

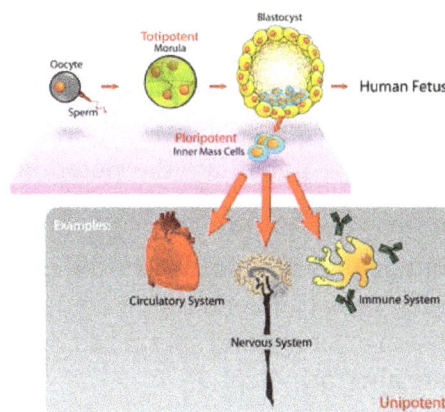

Pluripotent, embryonic stem cells originate as inner cell mass (ICM) cells within a blastocyst. These stem cells can become any tissue in the body, excluding a placenta. Only cells from an earli-er stage of the embryo, known as the morula, are totipotent, able to become all tissues in the body and the extraembryonic placenta.

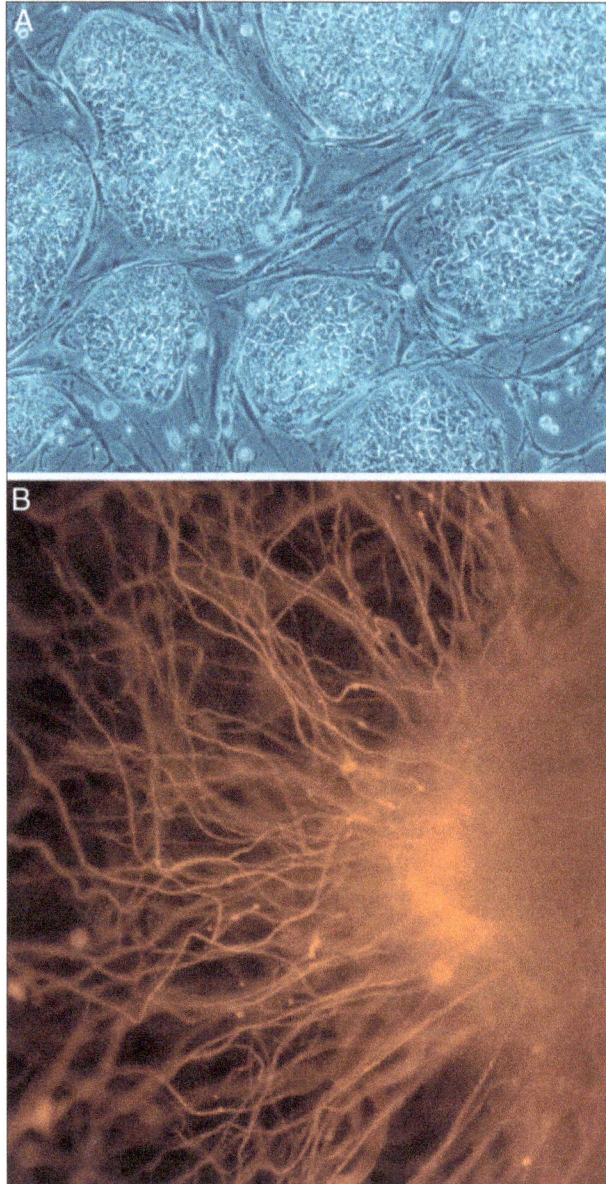

Human embryonic stem cells
A: Stem cell colonies that are not yet differentiated.
B: Nerve cells, an example of a cell type after differentiation.

Potency specifies the differentiation potential (the potential to differentiate into different cell types) of the stem cell.

- Totipotent (a.k.a. omnipotent) stem cells can differentiate into embryonic and extraembryonic cell types. Such cells can construct a complete, viable organism. These cells are produced from the fusion of an egg and sperm cell. Cells produced by the first few divisions of the fertilized egg are also totipotent.

- Pluripotent stem cells are the descendants of totipotent cells and can differentiate into nearly all cells, i.e. cells derived from any of the three germ layers.

- Multipotent stem cells can differentiate into a number of cell types, but only those of a closely related family of cells.

- Oligopotent stem cells can differentiate into only a few cell types, such as lymphoid or myeloid stem cells.

- Unipotent cells can produce only one cell type, their own, but have the property of self-renewal, which distinguishes them from non-stem cells (e.g. progenitor cells, which cannot self-renew).

Identification

In practice, stem cells are identified by whether they can regenerate tissue. For example, the defining test for bone marrow or hematopoietic stem cells (HSCs) is the ability to transplant the cells and save an individual without HSCs. This demonstrates that the cells can produce new blood cells over a long term. It should also be possible to isolate stem cells from the transplanted individual, which can themselves be transplanted into another individual without HSCs, demonstrating that the stem cell was able to self-renew.

Properties of stem cells can be illustrated *in vitro*, using methods such as clonogenic assays, in which single cells are assessed for their ability to differentiate and self-renew. Stem cells can also be isolated by their possession of a distinctive set of cell surface markers. However, *in vitro* culture conditions can alter the behavior of cells, making it unclear whether the cells will behave in a similar manner *in vivo*. There is considerable debate as to whether some proposed adult cell populations are truly stem cells.

Embryonic

Embryonic stem (ES) cells are the cells of the inner cell mass of a blastocyst, an early-stage embryo. Human embryos reach the blastocyst stage 4–5 days post fertilization, at which time they consist of 50–150 cells. ES cells are pluripotent and give rise during development to all derivatives of the three primary germ layers: ectoderm, endoderm and mesoderm. In other words, they can develop into each of the more than 200 cell types of the adult body when given sufficient and necessary stimulation for a specific cell type. They do not contribute to the extra-embryonic membranes or the placenta.

During embryonic development these inner cell mass cells continuously divide and become more specialized. For example, a portion of the ectoderm in the dorsal part of the embryo specializes as 'neurectoderm', which will become the future central nervous system. Later in development, neurulation causes the neurectoderm to form the neural tube. At the neural tube stage, the anterior portion undergoes encephalization to generate or 'pattern' the basic form of the brain. At this stage of development, the principal cell type of the CNS is considered a neural stem cell. These neural stem cells are pluripotent, as they can generate a large diversity of many different neuron types, each with unique gene expression, morphological, and functional characteristics. One prominent example of a neural stem cell is the radial glial cell, so named because it has a distinctive bipolar morphology with highly elongated processes spanning the thickness of the neural tube wall, and because historically it shared some glial characteristics, most notably the expression of glial fibrillary acidic protein (GFAP). The radial glial cell is the primary neural stem cell of the developing vertebrate CNS, and its cell body resides in the ventricular zone, adjacent to the developing ventricular system. Neural stem cells are committed to the neuronal lineages (neurons, astrocytes, and oligodendrocytes), and thus their potency is restricted.

Nearly all research to date has made use of mouse embryonic stem cells (mES) or human embryonic stem cells (hES) derived from the early inner cell mass. Both have the essential stem cell characteristics, yet they require very different environments in order to maintain an undifferentiated state. Mouse ES cells are grown on a layer of gelatin as an extracellular matrix (for support) and require the presence of leukemia inhibitory factor (LIF). Human ES cells are grown on a feeder layer of mouse embryonic fibroblasts (MEFs) and require the presence of basic fibroblast growth factor (bFGF or FGF-2). Without optimal culture conditions or genetic manipulation, embryonic stem cells will rapidly differentiate.

A human embryonic stem cell is also defined by the expression of several transcription factors and cell surface proteins. The transcription factors Oct-4, Nanog, and Sox2 form the core regulatory network that ensures the suppression of genes that lead to differentiation and the maintenance of pluripotency. The cell surface antigens most commonly used to identify hES cells are the glycolipids stage specific embryonic antigen 3 and 4 and the keratan sulfate antigens Tra-1-60 and Tra-1-81. By using human embryonic stem cells to produce specialized cells like nerve cells or heart cells in the lab, scientists can gain access to adult human cells without taking tissue from patients. They can then study these specialized adult cells in detail to try and catch complications of diseases, or to study cells reactions to potentially new drugs. The molecular definition of a stem cell includes many more proteins and continues to be a topic of research.

There are currently no approved treatments using embryonic stem cells. The first human trial was approved by the US Food and Drug Administration in January 2009. However, the human trial was not initiated until October 13, 2010 in Atlanta for spinal cord injury research. On November 14, 2011 the company conducting the trial (Geron Corporation) announced that it will discontinue further development of its stem cell programs. ES cells, being pluripotent cells, require specific signals for correct differentiation—if injected directly into another body, ES cells will differentiate into many different types of cells, causing a teratoma. Differentiating ES cells into usable cells while avoiding transplant rejection are just a few of the hurdles that embryonic stem cell researchers still face. Due to ethical considerations, many nations currently have moratoria or limitations on either human ES cell research or the production of new human ES cell lines. Because of their combined abilities of unlimited expansion and pluripotency, embryonic stem cells remain a theoretically potential source for regenerative medicine and tissue replacement after injury or disease.

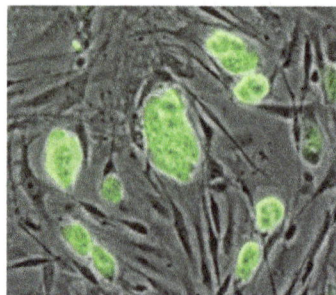

Mouse embryonic stem cells with fluorescent marker

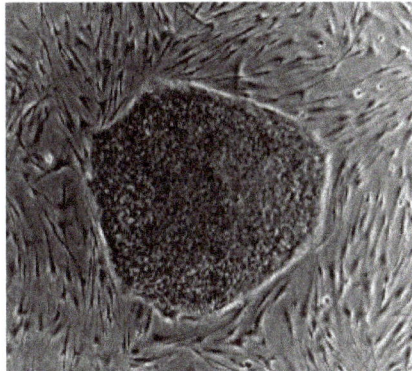

Human embryonic stem cell colony on mouse embryonic fibroblast feeder layer

Fetal

The primitive stem cells located in the organs of fetuses are referred to as fetal stem cells. There are two types of fetal stem cells:

1. Fetal proper stem cells come from the tissue of the fetus proper, and are generally obtained after an abortion. These stem cells are not immortal but have a high level of division and are multipotent.

2. Extraembryonic fetal stem cells come from extraembryonic membranes, and are generally not distinguished from adult stem cells. These stem cells are acquired after birth, they are not immortal but have a high level of cell division, and are pluripotent.

Adult

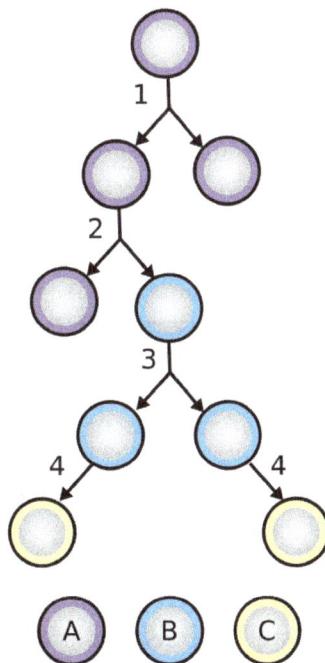

Stem cell division and differentiation. A: stem cell; B: progenitor cell; C: differentiated cell; 1: symmetric stem cell division; 2: asymmetric stem cell division; 3: progenitor division; 4: terminal differentiation

Adult stem cells, also called somatic (from Greek σωματικός, "of the body") stem cells, are stem cells which maintain and repair the tissue in which they are found. They can be found in children, as well as adults.

Pluripotent adult stem cells are rare and generally small in number, but they can be found in umbilical cord blood and other tissues. Bone marrow is a rich source of adult stem cells, which have been used in treating several conditions including liver cirrhosis, chronic limb ischemia and endstage heart failure. The quantity of bone marrow stem cells declines with age and is greater in males than females during reproductive years. Much adult stem cell research to date has aimed to characterize their potency and self-renewal capabilities. DNA damage accumulates with age in both stem cells and the cells that comprise the stem cell environment. This accumulation is con-sidered to be responsible, at least in part, for increasing stem cell dysfunction with aging.

Most adult stem cells are lineage-restricted (multipotent) and are generally referred to by their tissue origin (mesenchymal stem cell, adipose-derived stem cell, endothelial stem cell, dental pulp stem cell, etc.).

Adult stem cell treatments have been successfully used for many years to treat leukemia and related bone/blood cancers through bone marrow transplants. Adult stem cells are also used in veterinary medicine to treat tendon and ligament injuries in horses.

The use of adult stem cells in research and therapy is not as controversial as the use of embryonic stem cells, because the production of adult stem cells does not require the destruction of an embryo. Additionally, in instances where adult stem cells are obtained from the intended recipient (an autograft), the risk of rejection is essentially non-existent. Consequently, more US government funding is being provided for adult stem cell research.

Amniotic

Multipotent stem cells are also found in amniotic fluid. These stem cells are very active, expand extensively without feeders and are not tumorigenic. Amniotic stem cells are multipotent and can differentiate in cells of adipogenic, osteogenic, myogenic, endothelial, hepatic and also neuronal lines. Amniotic stem cells are a topic of active research.

Use of stem cells from amniotic fluid overcomes the ethical objections to using human embryos as a source of cells. Roman Catholic teaching forbids the use of embryonic stem cells in experimentation; accordingly, the Vatican newspaper "Osservatore Romano" called amniotic stem cells "the future of medicine".

It is possible to collect amniotic stem cells for donors or for autologuous use: the first US amniotic stem cells bank was opened in 2009 in Medford, MA, by Biocell Center Corporation and collaborates with various hospitals and universities all over the world.

Induced Pluripotent

These are not adult stem cells, but rather adult cells (e.g. epithelial cells) reprogrammed to give rise to pluripotent capabilities. Using genetic reprogramming with protein transcription factors, pluripotent stem cells equivalent to embryonic stem cells have been derived from human adult skin tissue. Shinya Yamanaka and his colleagues at Kyoto University used the transcription factors Oct3/4, Sox2, c-Myc, and Klf4 in their experiments on human facial skin cells. Junying Yu, James Thomson, and their colleagues at the University of Wisconsin–Madison used a different set of factors, Oct4, Sox2, Nanog and Lin28, and carried out their experiments using cells from human foreskin.

As a result of the success of these experiments, Ian Wilmut, who helped create the first cloned animal Dolly the Sheep, has announced that he will abandon somatic cell nuclear transfer as an avenue of research.

Frozen blood samples can be used as a source of induced pluripotent stem cells, opening a new avenue for obtaining the valued cells.

Lineage

To ensure self-renewal, stem cells undergo two types of cell division. Symmetric division gives rise to two identical daughter cells both endowed with stem cell properties. Asymmetric division, on the other hand, produces only one stem cell and a progenitor cell with limited self-renewal potential. Progenitors can go through several rounds of cell division before terminally differentiating into a mature cell. It is possible that the molecular distinction between symmetric and asymmetric divisions lies in differential segregation of cell membrane proteins (such as receptors) between the daughter cells.

An alternative theory is that stem cells remain undifferentiated due to environmental cues in their particular niche. Stem cells differentiate when they leave that niche or no longer receive those signals. Studies in *Drosophila* germarium have identified the signals decapentaplegic and adherens junctions that prevent germarium stem cells from differentiating.

Treatments

Stem cell therapy is the use of stem cells to treat or prevent a disease or condition. Bone marrow transplant is a form of stem cell therapy that has been used for many years without controversy. No stem cell therapies other than bone marrow transplant are widely used.

Disadvantages

Stem cell treatments may require immunosuppression because of a requirement for radiation before the transplant to remove the person's previous cells, or because the patient's immune system may target the stem cells. One approach to avoid the second possibility is to use stem cells from the same patient who is being treated.

Pluripotency in certain stem cells could also make it difficult to obtain a specific cell type. It is also difficult to obtain the exact cell type needed, because not all cells in a population differentiate uniformly. Undifferentiated cells can create tissues other than desired types.

Some stem cells form tumors after transplantation; pluripotency is linked to tumor formation especially in embryonic stem cells, fetal proper stem cells, induced pluripotent stem cells. Fetal proper stem cells form tumors despite multipotency.

Research

Some of the fundamental patents covering human embryonic stem cells are owned by the Wisconsin Alumni Research Foundation (WARF) - they are patents 5,843,780, 6,200,806, and 7,029,913 invented by James A. Thomson. WARF does not enforce these patents against academic scientists, but does enforce them against companies.

In 2006, a request for the US Patent and Trademark Office (USPTO) to re-examine the three patents was filed by the Public Patent Foundation on behalf of its client, the non-profit patent-watchdog group Consumer Watchdog (formerly the Foundation for Taxpayer and Consumer Rights). In the re-examination process, which involves several rounds of discussion between the USTPO and the parties, the USPTO initially agreed with Consumer Watchdog and rejected all the claims in all three patents, however in response, WARF amended the claims of all three patents to make them more narrow, and in 2008 the USPTO found the amended claims in all three patents to be patentable. The decision on one of the patents (7,029,913) was appealable, while the decisions on the other two were not. Consumer Watchdog appealed the granting of the '913 patent to the USTPO's Board of Patent Appeals and Interferences (BPAI) which granted the appeal, and in 2010 the BPAI decided that the amended claims of the '913 patent were not patentable. However, WARF was able to re-open prosecution of the case and did so, amending the claims of the '913 patent again to make them more narrow, and in January 2013 the amended claims were allowed.

In July 2013, Consumer Watchdog announced that it would appeal the decision to allow the claims of the '913 patent to the US Court of Appeals for the Federal Circuit (CAFC), the federal appeals court that hears patent cases. At a hearing in December 2013, the CAFC raised the question of whether Consumer Watchdog had legal standing to appeal; the case could not proceed until that issue was resolved.

Treatment

Potential uses of
Stem cells

Stroke
Traumatic brain injury
Learning defects
Alzheimer's disease
Parkinson's disease

Baldness
Blindness
Deafness

Missing teeth

Amyotrophic lateral-sclerosis

Wound healing

Bone marrow transplantation (currently established)

Myocardial infarction

Muscular dystrophy

Spinal cord injury

Diabetes

Osteoarthritis
Rheumatoid arthritis

Crohn's disease

Multiple sites:
Cancers

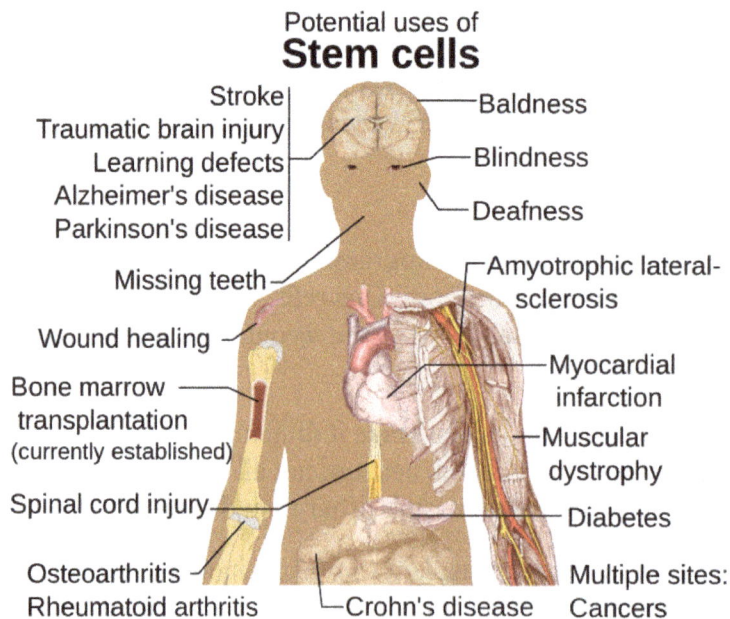

Diseases and conditions where stem cell treatment is being investigated.

Diseases and conditions where stem cell treatment is being investigated include:

- Diabetes
- Rheumatoid arthritis
- Parkinson's disease
- Alzheimer's disease
- Osteoarthritis
- Stroke and traumatic brain injury repair
- Learning disability due to congenital disorder
- Spinal cord injury repair
- Heart infarction
- Anti-cancer treatments
- Baldness reversal
- Replace missing teeth
- Repair hearing
- Restore vision and repair damage to the cornea
- Amyotrophic lateral sclerosis
- Crohn's disease
- Wound healing
- Male infertility due to absence of spermatogonial stem cells

Research is underway to develop various sources for stem cells, and to apply stem cell treatments for neurodegenerative diseases and conditions, diabetes, heart disease, and other conditions.

In more recent years, with the ability of scientists to isolate and culture embryonic stem cells, and with scientists' growing ability to create stem cells using somatic cell nuclear transfer and techniques to create induced pluripotent stem cells, controversy has crept in, both related to abortion politics and to human cloning.

Hepatotoxicity and drug-induced liver injury account for a substantial number of failures of new drugs in development and market withdrawal, highlighting the need for screening assays such as stem cell-derived hepatocyte-like cells, that are capable of detecting toxicity early in the drug development process.

Biomimetics

Biomimetics or biomimicry is the imitation of the models, systems, and elements of nature for the purpose of solving complex human problems. The terms "biomimetics" and "biomimicry" derive from Ancient Greek: βίος (*bios*), life, and μίμησις (*mīmēsis*), imitation, from μιμεῖσθαι (*mīmeisthai*), to imitate, from μῖμος (*mimos*), actor. A closely related field is bionics.

Velcro tape mimics biological examples of multiple hooked structures such as burs.

Living organisms have evolved well-adapted structures and materials over geological time through natural selection. Biomimetics has given rise to new technologies inspired by biological solutions at macro and nanoscales. Humans have looked at nature for answers to problems throughout our existence. Nature has solved engineering problems such as self-healing abilities, environmental exposure tolerance and resistance, hydrophobicity, self-assembly, and harnessing solar energy.

History

One of the early examples of biomimicry was the study of birds to enable human flight. Although never successful in creating a "flying machine", Leonardo da Vinci (1452–1519) was a keen observer of the anatomy and flight of birds, and made numerous notes and sketches on his observations as well as sketches of "flying machines". The Wright Brothers, who succeeded in flying the first heavier-than-air aircraft in 1903, derived inspiration from observations of pigeons in flight.

Biomimetics was coined by the American biophysicist and polymath Otto Schmitt during the 1950s. It was during his doctoral research that he developed the Schmitt trigger by studying the nerves in squid, attempting to engineer a device that replicated the biological system of nerve propagation. He continued to focus on devices that mimic natural systems and by 1957 he had perceived a converse to the standard view of biophysics at that time, a view he would come to call biomimetics.

Biophysics is not so much a subject matter as it is a point of view. It is an approach to problems of biological science utilizing the theory and technology of the physical sciences. Conversely, biophysics is also a biologist's approach to problems of physical science and engineering, although this aspect has largely been neglected.

— *Otto Herbert Schmitt, In Appreciation, A Lifetime of Connections: Otto Herbert Schmitt, 1913 - 1998*

A similar term, *Bionics* was coined by Jack E. Steele in 1960 at Wright-Patterson Air Force Base in Dayton, Ohio where Otto Schmitt also worked. Steele defined bionics as "the science of systems which have some function copied from nature, or which represent characteristics of natural systems or their analogues". During a later meeting in 1963 Schmitt stated,

Let us consider what bionics has come to mean operationally and what it or some word like it (I prefer biomimetics) ought to mean in order to make good use of the technical skills of scientists specializing, or rather, I should say, despecializing into this area of research

—*Otto Herbert Schmitt, In Appreciation, A Lifetime of Connections: Otto Herbert Schmitt, 1913 - 1998*

Velcro was inspired by the tiny hooks found on the surface of burs.

In 1969 the term biomimetics was used by Schmitt to title one of his papers, and by 1974 it had found its way into Webster's Dictionary, bionics entered the same dictionary earlier in 1960 as "a science concerned with the application of data about the functioning of biological systems to the solution of engineering problems". Bionic took on a different connotation when Martin Caidin referenced Jack Steele and his work in the novel *Cyborg* which later resulted in the 1974 television series *The Six Million Dollar Man* and its spin-offs. The term bionic then became associated with "the use of electronically operated artificial body parts" and "having ordinary human powers increased by or as if by the aid of such devices". Because the term *bionic* took on the implication of supernatural strength, the scientific community in English speaking countries largely abandoned it.

The term *biomimicry* appeared as early as 1982. Biomimicry was popularized by scientist and author Janine Benyus in her 1997 book *Biomimicry: Innovation Inspired by Nature*. Biomimicry is defined in the book as a "new science that studies nature's models and then imitates or takes inspiration from these designs and processes to solve human problems". Benyus suggests looking to Nature as a "Model, Measure, and Mentor" and emphasizes sustainability as an objective of biomimicry.

Existing Commercialized Applications

Fabrication

Biomorphic mineralization is a technique that produces materials with morphologies and structures resembling those of natural living organisms by using bio-structures as templates for mineralization. Compared to other methods of material production, biomorphic mineralization is facile, environmentally benign and economic.

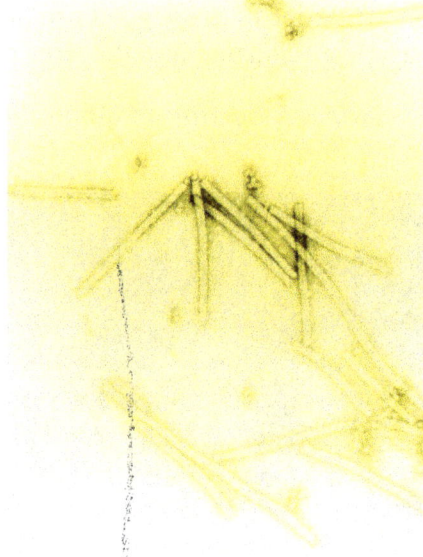

Scanning electron micrograph of rod shaped tobacco mosaic virus particles

Display Technology

Vibrant blue color of *Morpho* butterfly due to structural coloration

Morpho butterfly wings contain microstructures that create its coloring effect through structural coloration rather than pigmentation. Incident light waves are reflected at specific wavelengths to create vibrant colors due to multilayer interference, diffraction, thin film interference, and scattering properties. The scales of these butterflies consist of microstructures such as ridges, cross-ribs, ridge-lamellae, and microribs that have been shown to be responsible for coloration. The structural color has been simply explained as the interference due to alternating layers of cuticle and air using a model of multilayer interference. The same principles behind the coloration of soap bubbles apply to butterfly wings. The color of butterfly wings is due to multiple instances of constructive interference from structures such as this. The photonic microstructure of butterfly wings can be replicated through biomorphic mineralization to yield similar properties. The photonic microstructures can be replicated using metal oxides or metal alkoxides such as titanium sulfate ($TiSO_4$), zirconium oxide (ZrO_2), and aluminium oxide (Al_2O_3). An alternative method of vapor-phase oxidation of SiH4 on the template surface was found to preserve delicate structural features of the microstructure. A display technology ("Mirasol") based on the reflective properties of *Morpho* butterfly wings was commercialized by Qualcomm in 2007. The technology uses Interferometric Modulation to reflect light so only the desired color is visible in each individual pixel of the display.

Possible Future Applications

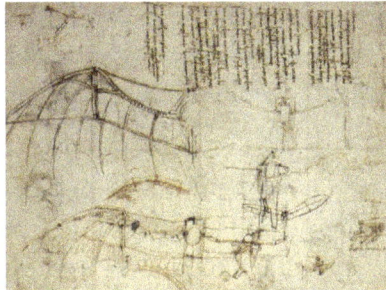

Leonardo da Vinci's design for a flying machine with wings based closely upon the structure of bat wings

Biomimetics could in principle be applied in many fields. Because of the complexity of biological systems, the number of features that might be imitated is large. Biomimetic applications are at various stages of development from technologies that might become commercially usable to prototypes.

Prototypes

Researchers studied the termite's ability to maintain virtually constant temperature and humidity in their termite mounds in Africa despite outside temperatures that vary from 1.5 °C to 40 °C (35 °F to 104 °F). Researchers initially scanned a termite mound and created 3-D images of the mound structure, which revealed construction that could influence human building design. The Eastgate Centre, a mid-rise office complex in Harare, Zimbabwe, stays cool without air conditioning and uses only 10% of the energy of a conventional building of the same size.

In structural engineering, the Swiss Federal Institute of Technology (EPFL) has incorporated biomimetic characteristics in an adaptive deployable "tensegrity" bridge. The bridge can carry out self-diagnosis and self-repair.

Technologies

Practical underwater adhesion is an engineering challenge since current technology is unable to stick surface strongly underwater because of barriers such as hydration layers and contaminants on surfaces. However, marine mussels can stick easily and efficiently to surfaces underwater under the harsh conditions of the ocean. They use strong filaments to adhere to rocks in the inter-tidal zones of wave-swept beaches, preventing them from being swept away in strong sea currents. Mussel foot proteins attach the filaments to rocks, boats and practically any surface in nature including other mussels. These proteins contain a mix of amino acid residues which has been adapted specifically for adhesive purposes. Researchers from the University of California Santa Barbara borrowed and simplified chemistries that the mussel foot uses to overcome this engineering challenge of wet adhesion to create copolyampholytes, and one-component adhesive systems with potential for employment in nanofabrication protocols.

Mimicking the diving behavior of animals, researchers discovered in 2013 that humans have a similar capacity to lower brain temperature and suppress metabolism for neuroprotection. This has now opened a real possibility of devising means for humans to sustain this state, not unlike the elusive and enigmatic feat of animal hibernation, e.g., lemurs (primates) and bears. This would have profound biomedical implications for healthcare and for treating an unmatched range and diversity of serious life-threatening clinical conditions, and in a fully personalized way, things like stroke, blood-loss, burns, cancer, chronic obesity, epileptic seizures, etc. An experimental trial, recently conducted in Sweden seemingly resulted in a sustainable variant of this state in a human breath-hold diver.

Spider web silk is as strong as the Kevlar used in bulletproof vests. Engineers could in principle use such a material, if it could be reengineered to have a long enough life, for parachute lines, suspension bridge cables, artificial ligaments for medicine, and other purposes. Other research has proposed adhesive glue from mussels, solar cells made like leaves, fabric that emulates shark skin, harvesting water from fog like a beetle, and more. Murray's law, which in conventional form determined the optimum diameter of blood vessels, has been re-derived to provide simple equations for the pipe or tube diameter which gives a minimum mass engineering system. Aircraft wing design and flight techniques are being inspired by birds and bats.

Robots based on the physiology and methods of locomotion of animals include BionicKangaroo which moves like a kangaroo, saving energy from one jump and transferring it to its next jump, and climbing robots, boots and tape mimicking geckos feet and their ability for adhesive reversal. Nanotechnology surfaces that recreate properties of shark skin are intended to enable more efficient movement through water. Tire treads have been inspired by the toe pads of tree frogs. The self-sharpening teeth of many animals have been copied to make better cutting tools. Protein folding is used to control material formation for self-assembled functional nanostructures. The Structural coloration of butterfly wings is adapted to provide improved interferometric modulator displays and everlasting colours. New ceramics copy the properties of seashells. Polar bear fur has inspired the design of thermal collectors and clothing. The arrangement of leaves on a plant has been adapted for better solar power collection. The light refractive properties of the moth's eye has been studied to reduce the reflectivity of solar panels. Self-healing materials, polymers and composite materials capable of mending cracks have been produced based on biological materials.

The Bombardier beetle's powerful repellent spray inspired a Swedish company to develop a "micro mist" spray technology, which is claimed to have a low carbon impact (compared to aerosol sprays). The beetle mixes chemicals and releases its spray via a steerable nozzle at the end of its abdomen, stinging and confusing the victim.

Most viruses have an outer capsule 20 to 300 nm in diameter. Virus capsules are remarkably robust and capable of withstanding temperatures as high as 60 °C; they are stable across the pH range 2-10. Viral capsules can be used to create nano device components such as nanowires, nanotubes, and quantum dots. Tubular virus particles such as the tobacco mosaic virus (TMV) can be used as templates to create nanofibers and nanotubes, since both the inner and outer layers of the virus are charged surfaces which can induce nucleation of crystal growth. This was demonstrated through the production of platinum and gold nanotubes using TMV as a template. Mineralized virus particles have been shown to withstand various pH values by mineralizing the viruses with different materials such as silicon, PbS, and CdS and could therefore serve as a useful carriers of material. A spherical plant virus called cowpea chlorotic mottle virus (CCMV) has interesting expanding properties when exposed to environments of pH higher than 6.5. Above this pH, 60 independent pores with diameters about 2 nm begin to exchange substance with the environment. The structural transition of the viral capsid can be utilized in Biomorphic mineralization for selective uptake and deposition of minerals by controlling the solution pH. Possible applications include using the viral cage to produce uniformly shaped and sized quantum dot semiconductor nanoparticles through a series of pH washes. This is an alternative to the apoferritin cage technique currently used to synthesize uniform CdSe nanoparticles. Such materials could also be used for targeted drug delivery since particles release contents upon exposure to specific pH levels.

Tissue Culture

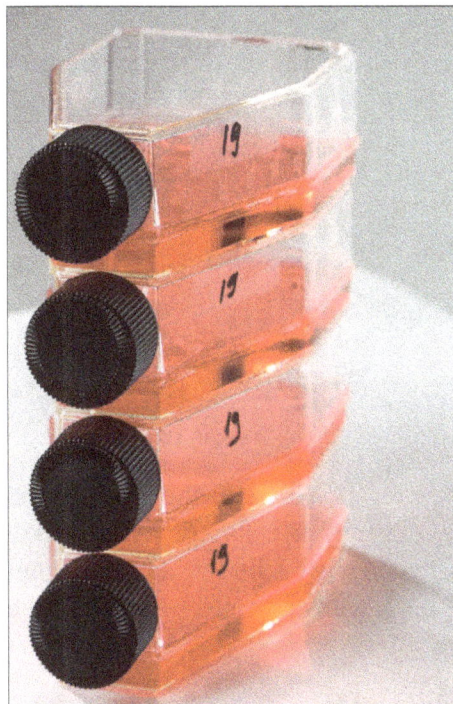

Flasks containing tissue culture growth medium which provides nourishment to growing cells.

Tissue culture is the growth of tissues or cells separate from the organism. This is typically facilitated via use of a liquid, semi-solid, or solid growth medium, such as broth or agar. Tissue culture commonly refers to the culture of animal cells and tissues, with the more specific term plant tissue culture being used for plants. The term "tissue culture" was coined by American pathologist Montrose Thomas Burrows, M.D.

Historical Usage

In 1885 Wilhelm Roux removed a section of the medullary plate of an embryonic chicken and maintained it in a warm saline solution for several days, establishing the basic principle of tissue culture. In 1907 the zoologist Ross Granville Harrison demonstrated the growth of frog embryonic cells that would give rise to nerve cells in a medium of clotted lymph. In 1913, E. Steinhardt, C. Israeli, and R. A. Lambert grew vaccinia virus in fragments of guinea pig corneal tissue. In 1996, the first use of regenerative tissue was used to replace a small length of urethra, which led to the understanding that the technique of obtaining samples of tissue, growing it outside the body without a scaffold, and reapplying it, can be used for only small distances of less than 1 cm.

Gottlieb Haberlandt first pointed out the possibilities of the culture of isolated tissues, plant tissue culture. He suggested that the potentialities of individual cells via tissue culture as well as that the reciprocal influences of tissues on one another could be determined by this method. Since Haberlandt's original assertions, methods for tissue and cell culture have been realized, leading to significant discoveries in biology and medicine. His original idea, presented in 1902, was called totipotentiality: "Theoretically all plant cells are able to give rise to a complete plant."

Modern Usage

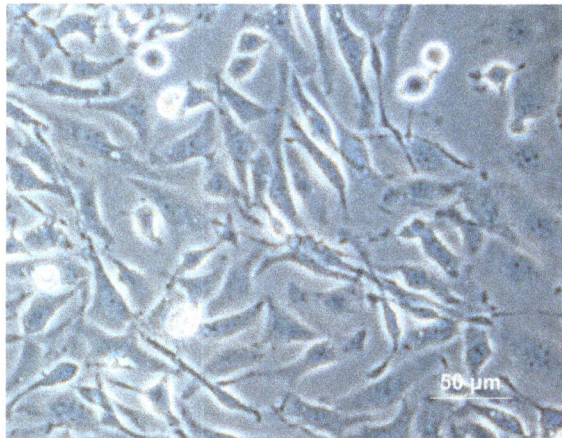

Cultured cells growing in growth medium

In modern usage, tissue culture generally refers to the growth of cells from a tissue from a multicellular organism *in vitro*. These cells may be cells isolated from a donor organism, "primary cells", or an immortalised cell line. The cells are bathed in a culture medium, which contains essential nutrients and energy sources necessary for the cells' survival. The term tissue culture is often used interchangeably with cell culture

The literal meaning of tissue culture refers to the culturing of tissue pieces, i.e. explant culture.

Tissue culture is an important tool for the study of the biology of cells from multicellular organisms. It provides an *in vitro* model of the tissue in a well defined environment which can be easily manipulated and analysed.

Plant tissue culture in particular is concerned with the growing of entire plants from small pieces of plant tissue, cultured in medium

References

- Kilusang Magbubukid ng Pilipinas (2007). Victoria M. Lopez; et al., eds. The Great Riice Robbery: A Handbook on the Impact of IRRI in Asia (PDF). Penang, Malaysia: Pesticide Action Network Asia and the Pacific. ISBN 978-983-9381-35-1. Retrieved 8 August 2011.

- Ponting, Clive (2007). A New Green History of the World: The Environment and the Collapse of Great Civilizations. New York: Penguin Books. p. 244. ISBN 978-0-14-303898-6.

- Schöler, Hans R. (2007). "The Potential of Stem Cells: An Inventory". In Nikolaus Knoepffler; Dagmar Schipanski; Stefan Lorenz Sorgner. Humanbiotechnology as Social Challenge. Ashgate Publishing. p. 28. ISBN 978-0-7546-5755-2.

- Gilbert, Scott F.; College, Swarthmore; Helsinki, the University of (2014). Developmental biology (Tenth ed.). Sunderland, Mass.: Sinauer. ISBN 978-0878939787.

- Ariff Bongso; Eng Hin Lee, eds. (2005). "Stem cells: their definition, classification and sources". Stem Cells: From Benchtop to Bedside. World Scientific. p. 5. ISBN 981-256-126-9. OCLC 443407924.

- Hanna V, Gassei K, Orwig KE (2015). "Stem Cell Therapies for Male Infertility: Where Are We Now and Where Are We Going?". Biennial Review of Infertility. Bone marrow transplantation is, as of 2009, the only established use of stem cells.

- Vincent, Julian F. V.; et al. (22 August 2006). "Biomimetics: its practice and theory". doi:10.1098/rsif.2006.0127. Retrieved 7 April 2015.

- Cathey, Jim (7 January 2010). "Nature Knows Best: What Burrs, Geckos and Termites Teach Us About Design". Qualcomm. Retrieved 24 August 2015.

- Ackerman, Evan (2 Apr 2014). "Festo's Newest Robot Is a Hopping Bionic Kangaroo". spectrum.ieee.org. IEEE Spectrum. Retrieved 17 Apr 2014.

- "The Secret of the Fibonacci Sequence in Trees". 2011 Winning Essays. American Museum of Natural History. 1 May 2014. Retrieved 17 July 2014.

- Behrens A, van Deursen JM, Rudolph KL, Schumacher B (2014). "Impact of genomic damage and ageing on stem cell function". Nat. Cell Biol. 16 (3): 201–7. doi:10.1038/ncb2928. PMC 4214082. PMID 24576896.

- "Green Revolution research saved an estimated 18 to 27 million hectares from being brought into agricultural production". Pnas.org. 2013-05-13. Retrieved 2013-08-28.

- Ball, Philip (May 2012). "Scientific American". Nature's Color Tricks. 306. pp. 74–79. doi:10.1038/scientificamerican0512-74. Retrieved 3 June 2012.

3

Processes of Biotechnology

Selective breeding is the process by which humans breed animals and plants with specific characteristics. This chapter discusses the process of selective breeding, biosynthesis, catalysis and hybrid. It explains in detail the process of selective breeding and hybrid breeding.

Selective Breeding

A Belgian Blue cow. The defect in the breed's myostatin gene is maintained through linebreeding and is responsible for its accelerated lean muscle growth.

Selective breeding (also called artificial selection) is the process by which humans use animal breeding and plant breeding to selectively develop particular phenotypic traits (characteristics) by choosing which typically animal or plant males and females will sexually reproduce and have offspring together. Domesticated animals are known as breeds, normally bred by a professional breeder, while domesticated plants are known as varieties, cultigens, or cultivars. Two purebred animals of different breeds produce a crossbreed, and crossbred plants are called hybrids. Flowers, vegetables and fruit-trees may be bred by amateurs and commercial or non-commercial professionals: major crops are usually the provenance of the professionals.

There are two approaches or types of artificial selection, or selective breeding. First is the traditional "breeder's approach" in which the breeder or experimenter applies "a known amount of selection to a single phenotypic trait" by examining the chosen trait and choosing to breed only those that exhibit higher or "extreme values" of that trait. The second is called "controlled natural selection," which is essentially natural selection in a controlled environment. In this, the breeder does not choose which individuals being tested "survive or reproduce," as he or she could in the traditional approach. There are also "selection experiments," which is a third approach and these are conducted in order to determine the "strength of natural selection in the wild." However, this is more often an observational approach as opposed to an experimental approach.

This Chihuahua mix and Great Dane shows the wide range of dog
breed sizes created using selective breeding.

In animal breeding, techniques such as inbreeding, linebreeding, and outcrossing are utilized. In plant breeding, similar methods are used. Charles Darwin discussed how selective breeding had been successful in producing change over time in his book, *On the Origin of Species*. The first chapter of the book discusses selective breeding and domestication of such animals as pigeons, cats, cattle, and dogs. Selective breeding was used by Darwin as a springboard to introduce the theory of natural selection, and to support it.

Selective breeding transformed teosinte's few fruitcases (left) into modern
maize's rows of exposed kernels (right).

The deliberate exploitation of selective breeding to produce desired results has become very common in agriculture and experimental biology.

Selective breeding can be unintentional, e.g., resulting from the process of human cultivation; and it may also produce unintended – desirable or undesirable – results. For example, in some grains, an increase in seed size may have resulted from certain ploughing practices rather than from the intentional selection of larger seeds. Most likely, there has been an interdependence between natural and artificial factors that have resulted in plant domestication.

History

Selective breeding of both plants and animals has been practiced since early prehistory; key species such as wheat, rice, and dogs have been significantly different from their wild ancestors for millennia, and maize, which required especially large changes from teosinte, its wild form, was selectively bred in Mesoamerica. Selective breeding was practiced by the Romans. Treatises as much as 2,000 years old give advice on selecting animals for different purposes, and these ancient works cite still older authorities, such as Mago the Carthaginian. The notion of selective breeding was later expressed by the Persian Muslim polymath Abu Rayhan Biruni in the 11th century. He noted the idea in his book titled *India*, and gave various examples.

The agriculturist selects his corn, letting grow as much as he requires, and tearing out the remainder. The forester leaves those branches which he perceives to be excellent, whilst he cuts away all others. The bees kill those of their kind who only eat, but do not work in their beehive.

Selective breeding was established as a scientific practice by Robert Bakewell during the British Agricultural Revolution in the 18th century. Arguably, his most important breeding program was with sheep. Using native stock, he was able to quickly select for large, yet fine-boned sheep, with long, lustrous wool. The Lincoln Longwool was improved by Bakewell, and in turn the Lincoln was used to develop the subsequent breed, named the New (or Dishley) Leicester. It was hornless and had a square, meaty body with straight top lines.

These sheep were exported widely, including to Australia and North America, and have contributed to numerous modern breeds, despite the fact that they fell quickly out of favor as market preferences in meat and textiles changed. Bloodlines of these original New Leicesters survive today as the English Leicester (or Leicester Longwool), which is primarily kept for wool production.

Bakewell was also the first to breed cattle to be used primarily for beef. Previously, cattle were first and foremost kept for pulling ploughs as oxen, but he crossed long-horned heifers and a Westmoreland bull to eventually create the Dishley Longhorn. As more and more farmers followed his lead, farm animals increased dramatically in size and quality. In 1700, the average weight of a bull sold for slaughter was 370 pounds (168 kg). By 1786, that weight had more than doubled to 840 pounds (381 kg). However, after his death, the Dishley Longhorn was replaced with short-horn versions.

He also bred the Improved Black Cart horse, which later became the Shire horse.

Charles Darwin coined the term 'selective breeding'; he was interested in the process as an illustration of his proposed wider process of natural selection. Darwin noted that many domesticated animals and plants had special properties that were developed by intentional animal and plant breeding from individuals that showed desirable characteristics, and discouraging the breeding of individuals with less desirable characteristics.

Darwin used the term "artificial selection" twice in the 1859 first edition of his work *On the Origin of Species*, in Chapter IV: Natural Selection, and in Chapter VI: Difficulties on Theory –

Slow though the process of selection may be, if feeble man can do much by his powers of artificial selection, I can see no limit to the amount of change, to the beauty and infinite complexity of the co-adaptations between all organic beings, one with another and with their physical conditions of life, which may be effected in the long course of time by nature's power of selection.

We are profoundly ignorant of the causes producing slight and unimportant variations; and we are immediately made conscious of this by reflecting on the differences in the breeds of our domesticated animals in different countries,—more especially in the less civilized countries where there has been but little artificial selection.

Animal Breeding

Three generations of "Westies" in a village in Fife, Scotland

Animals with homogeneous appearance, behavior, and other characteristics are known as particular breeds, and they are bred through culling animals with particular traits and selecting for further breeding those with other traits. Purebred animals have a single, recognizable breed, and purebreds with recorded lineage are called pedigreed. Crossbreeds are a mix of two purebreds, whereas mixed breeds are a mix of several breeds, often unknown. Animal breeding begins with breeding stock, a group of animals used for the purpose of planned breeding. When individuals are looking to breed animals, they look for certain valuable traits in purebred stock for a certain purpose, or may intend to use some type of crossbreeding to produce a new type of stock with different, and, it is presumed, superior abilities in a given area of endeavor. For example, to breed chickens, a typical breeder intends to receive eggs, meat, and new, young birds for further reproduction. Thus, the breeder has to study different breeds and types of chickens and analyze what can be expected from a certain set of characteristics before he or she starts breeding them. Therefore, when purchasing initial breeding stock, the breeder seeks a group of birds that will most closely fit the purpose intended.

Purebred breeding aims to establish and maintain stable traits, that animals will pass to the next generation. By "breeding the best to the best," employing a certain degree of inbreeding, considerable culling, and selection for "superior" qualities, one could develop a bloodline superior in certain respects to the original base stock. Such animals can be recorded with a breed registry, the organization that maintains pedigrees and/or stud books. However, single-trait breeding, breeding for only one trait over all others, can be problematic. In one case mentioned by animal behaviorist Temple Grandin, roosters bred for fast growth or heavy muscles did not know how to perform typical rooster courtship dances, which alienated the roosters from hens and led the roosters to kill the hens after reproducing with them.

The observable phenomenon of hybrid vigor stands in contrast to the notion of breed purity. However, on the other hand, indiscriminate breeding of crossbred or hybrid animals may also result in degradation of quality. Studies in evolutionary physiology, behavioral genetics, and other areas of organismal biology have also made use of deliberate selective breeding, though longer generation times and greater difficulty in breeding can make such projects challenging in vertebrates.

Plant Breeding

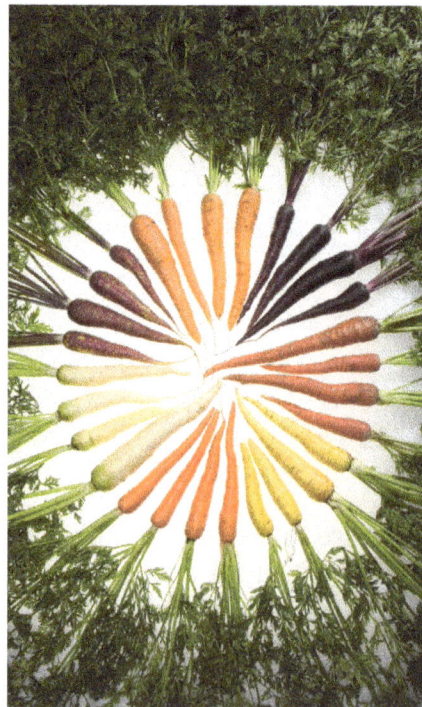

Researchers at the USDA have selectively bred carrots with a variety of colors.

Plant breeding has been used for thousands of years, and began with the domestication of wild plants into uniform and predictable agricultural cultigens. High-yielding varieties have been particularly important in agriculture.

Selective plant breeding is also used in research to produce transgenic animals that breed "true" (i.e., are homozygous) for artificially inserted or deleted genes.

Selective Breeding in Aquaculture

Selective breeding in aquaculture holds high potential for the genetic improvement of fish and shellfish. Unlike terrestrial livestock, the potential benefits of selective breeding in aquaculture were not realized until recently. This is because high mortality led to the selection of only a few broodstock, causing inbreeding depression, which then forced the use of wild broodstock. This was evident in selective breeding programs for growth rate, which resulted in slow growth and high mortality.

Control of the reproduction cycle was one of the main reasons as it is a requisite for selective breeding programmes. Artificial reproduction was not achieved because of the difficulties in hatching or feeding some farmed species such as eel and yellowtail farming. A suspected reason associated with the late realisation of success in selective breeding programs in aquaculture was the education of the concerned people – researchers, advisory personnel and fish farmers. The education of fish biologists paid less attention to quantitative genetics and breeding plans.

Another was the failure of documentation of the genetic gains in successive generations. This in turn led to failure in quantifying economic benefits that successful selective breeding programs produce. Documentation of the genetic changes was considered important as they help in fine tuning further selection schemes.

Quality Traits in Aquaculture

Aquaculture species are reared for particular traits such as growth rate, survival rate, meat quality, resistance to diseases, age at sexual maturation, fecundity, shell traits like shell size, shell colour, etc.

- Growth rate – growth rate is normally measured as either body weight or body length. This trait is of great economic importance for all aquaculture species as faster growth rate speeds up the turnover of production. Improved growth rates show that farmed animals utilize their feed more efficiently through a correlated response.

- Survival rate – survival rate may take into account the degrees of resistance to diseases. This may also see the stress response as fish under stress are highly vulnerable to diseases. The stress fish experience could be of biological, chemical or environmental influence.

- Meat quality – the quality of fish is of great economic importance in the market. Fish quality usually takes into account size, meatiness, and percentage of fat, colour of flesh, taste, shape of the body, ideal oil and omega-3 content.

- Age at sexual maturation – The age of maturity in aquaculture species is another very important attribute for farmers as during early maturation the species divert all their energy to gonad production affecting growth and meat production and are more susceptible to health problems (Gjerde 1986).

- Fecundity – As the fecundity in fish and shellfish is usually high it is not considered as a major trait for improvement. However, selective breeding practices may consider the size of the egg and correlate it with survival and early growth rate.

Finfish Response to Selection

Salmonids

Gjedrem (1979) showed that selection of Atlantic salmon (*Salmo salar*) led to an increase in body weight by 30% per generation. A comparative study on the performance of select Atlantic salmon with wild fish was conducted by AKVAFORSK Genetics Centre in Norway. The traits, for which the selection was done included growth rate, feed consumption, protein retention, energy retention, and feed conversion efficiency. Selected fish had a twice better growth rate, a 40% higher feed intake, and an increased protein and energy retention. This led to an overall 20% better Fed Conversion Efficiency as compared to the wild stock. Atlantic salmon have also been selected for resistance to bacterial and viral diseases. Selection was done to check resistance to Infectious Pancreatic Necrosis Virus (IPNV). The results showed 66.6% mortality for low-resistant species whereas the high-resistant species showed 29.3% mortality compared to wild species.

Rainbow trout (*S. gairdneri*) was reported to show large improvements in growth rate after 7–10 generations of selection. Kincaid et al. (1977) showed that growth gains by 30% could be achieved by selectively breeding rainbow trout for three generations. A 7% increase in growth was recorded per generation for rainbow trout by Kause et al. (2005).

In Japan, high resistance to IPNV in rainbow trout has been achieved by selectively breeding the stock. Resistant strains were found to have an average mortality of 4.3% whereas 96.1% mortality was observed in a highly sensitive strain.

Coho salmon (*Oncorhynchus kisutch*) increase in weight was found to be more than 60% after four generations of selective breeding. In Chile, Neira et al. (2006) conducted experiments on early spawning dates in coho salmon. After selectively breeding the fish for four generations, spawning dates were 13–15 days earlier.

Cyprinids

Selective breeding programs for the Common carp (*Cyprinus carpio*) include improvement in growth, shape and resistance to disease. Experiments carried out in the USSR used crossings of broodstocks to increase genetic diversity and then selected the species for traits like growth rate, exterior traits and viability, and/or adaptation to environmental conditions like variations in temperature. Kirpichnikov *et al.* (1974) and Babouchkine (1987) selected carp for fast growth and tolerance to cold, the Ropsha carp. The results showed a 30–40% to 77.4% improvement of cold tolerance but did not provide any data for growth rate. An increase in growth rate was observed in the second generation in Vietnam. Moav and Wohlfarth (1976) showed positive results when selecting for slower growth for three generations compared to selecting for faster growth. Schaperclaus (1962) showed resistance to the dropsy disease wherein selected lines suffered low mortality (11.5%) compared to unselected (57%).

Channel Catfish

Growth was seen to increase by 12–20% in selectively bred *Iictalurus punctatus*. More recently, the response of the Channel Catfish to selection for improved growth rate was found to be approximately 80%, i.e., an average of 13% per generation.

Shellfish Response to Selection

Oysters

Selection for live weight of Pacific oysters showed improvements ranging from 0.4% to 25.6% compared to the wild stock. Sydney-rock oysters (*Saccostrea commercialis*) showed a 4% increase after one generation and a 15% increase after two generations. Chilean oysters (*Ostrea chilensis*), selected for improvement in live weight and shell length showed a 10–13% gain in one generation. Bonamia ostrea is a protistan parasite that causes catastrophic losses (nearly 98%) in European flat oyster *Ostrea edulis* L. This protistan parasite is endemic to three oyster-regions in Europe. Selective breeding programs show that *O. edulis* susceptibility to the infection differs across oyster strains in Europe. A study carried out by Culloty et al. showed that 'Rossmore' oysters in Cork harbour, Ireland had better resistance compared to other Irish strains. A selective breeding program at Cork harbour uses broodstock from 3– to 4-year-old survivors and is further controlled until a viable percentage reaches market size. Over the years 'Rossmore' oysters have shown to develop lower prevalence to *B. ostreae* infection and percentage mortality. Ragone Calvo et al. (2003) selectively bred the eastern oyster, *Crassostrea virginica*, for resistance against co-occurring parasites *Haplosporidium nelson* (MSX) and *Perkinsus marinus* (Dermo). They achieved dual resistance to the disease in four generations of selective breeding. The oysters showed higher growth and survival rates and low susceptibility to the infections. At the end of the experiment, artificially selected *C. virginica* showed a 34–48% higher survival rate.

Penaeid Shrimps

Selection for growth in Penaeid shrimps yielded successful results. A selective breeding program for *Litopenaeus stylirostris* saw an 18% increase in growth after the fourth generation and 21% growth after the fifth generation. *Marsupenaeus japonicas* showed a 10.7% increase in growth after the first generation. Argue et al. (2002) conducted a selective breeding program on the Pacific White Shrimp, *Litopenaeus vannamei* at The Oceanic Institute, Waimanalo, USA from 1995 to 1998. They reported significant responses to selection compared to the unselected control shrimps. After one generation, a 21% increase was observed in growth and 18.4% increase in survival to TSV. The Taura Syndrome Virus (TSV) causes mortalities of 70% or more in shrimps. C.I. Oceanos S.A. in Colombia selected the survivors of the disease from infected ponds and used them as parents for the next generation. They achieved satisfying results in two or three generations wherein survival rates approached levels before the outbreak of the disease. The resulting heavy losses (up to 90%) caused by Infectious hypodermal and haematopoietic necrosis virus (IHHNV) caused a number of shrimp farming industries started to selectively breed shrimps resistant to this disease. Successful outcomes led to development of Super Shrimp, a selected line of *L. stylirostris* that is resistant to IHHNV infection. Tang et al. (2000) confirmed this by showing no mortalities in IHHNV- challenged Super Shrimp post larvae and juveniles.

Aquatic Species Versus Terrestrial Livestock

Selective breeding programs for aquatic species provide better outcomes compared to terrestrial livestock. This higher response to selection of aquatic farmed species can be attributed to the following:

- High fecundity in both sexes fish and shellfish enabling higher selection intensity.

- Large phenotypic and genetic variation in the selected traits.

Selective breeding in aquaculture provide remarkable economic benefits to the industry, the primary one being that it reduces production costs due to faster turnover rates. This is because of faster growth rates, decreased maintenance rates, increased energy and protein retention, and better feed efficiency. Applying such genetic improvement program to aquaculture species will increase productivity to meet the increasing demands of growing populations.

Advantages and Disadvantages

Selective breeding is a direct way to determine if a specific trait can "evolve in response to selection." A single-generation method of breeding is not as accurate or direct. The process is also more practical and easier to understand than sibling analysis. The former tests "differences between line means" while the latter is dependent upon "variance and covariance components." Essentially, selective breeding is better for traits such as physiology and behavior that are hard to measure because it requires fewer individuals to test than single-generation testing.

However, there are disadvantages to this process. Because a single experiment done in selective breeding cannot be used to assess an entire group of "genetic variances and covariances," individual experiments must be done for every individual trait. Also, because of the necessity of selective breeding experiments to require maintaining the organisms tested in a lab or greenhouse, it is impractical to use this breeding method on many organisms. Controlled mating instances are difficult to carry out in this case and this is a necessary component of selective breeding.

Hybrid (Biology)

Hercules, a "liger", a lion/tiger hybrid.

In biology a hybrid, also known as cross breed, is the result of mixing, through sexual reproduction, two animals or plants of different breeds, varieties, species or genera. Using genetic terminology, it may be defined as follows.

1. *Hybrid* generally refers to any offspring resulting from the breeding of two genetically distinct individuals, which usually will result in a high degree of heterozygosity, though hybrid and heterozygous are not, strictly speaking, synonymous.

2. A *genetic hybrid* carries two different alleles of the same gene

3. A *structural hybrid* results from the fusion of gametes that have differing structure in at least one chromosome, as a result of structural abnormalities

4. A *numerical hybrid* results from the fusion of gametes having different haploid numbers of chromosomes

5. A *permanent hybrid* is a situation where only the heterozygous genotype occurs, because all homozygous combinations are lethal.

From a taxonomic perspective, hybrid refers to:

1. Offspring resulting from the interbreeding between two animal species or plant species.

2. Hybrids between different subspecies within a species (such as between the Bengal tiger and Siberian tiger) are known as *intra-specific* hybrids. Hybrids between different species within the same genus (such as between lions and tigers) are sometimes known as *interspecific* hybrids or crosses. Hybrids between different genera (such as between sheep and goats) are known as *intergeneric* hybrids. Extremely rare *interfamilial* hybrids have been known to occur (such as the guineafowl hybrids). No *interordinal* (between different orders) animal hybrids are known.

3. The third type of hybrid consists of crosses between populations, breeds or cultivars within a single species. This meaning is often used in plant and animal breeding, where hybrids are commonly produced and selected, because they have desirable characteristics not found or inconsistently present in the parent individuals or populations.

Terminology

The term hybrid is derived from Latin *hybrida*, meaning the *"offspring of a tame sow and a wild boar"*, *"child of a freeman and slave"*, etc. The term came into popular use in English in the 19th century, though examples of its use have been found from the early 17th century.

There is a popular convention of naming hybrids by forming portmanteau words. The template for this is the naming of tiger-lion hybrids as liger and tigon in the 1920s. This was playfully (but unsystematically) extended to a number of other hybrids, or hypothetical hybrids, such as beefalo (1960s), humanzee (1980s), cama (1998).

Types of Hybrids

Depending on the parents, there are a number of different types of hybrids;

- *Single cross hybrids* — result from the cross between two true breeding organisms and produces an F1 generation called an F1 hybrid (F1 is short for Filial 1, meaning "first off-spring"). The cross between two different homozygous lines produces an F1 hybrid that is heterozygous; having two alleles, one contributed by each parent and typically one is dominant and the other recessive. Typically, the F1 generation is also phenotypically homogeneous, producing offspring that are all similar to each other.

- *Double cross hybrids* — result from the cross between two different F1 hybrids.

- *Three-way cross hybrids* — result from the cross between one parent that is an F1 hybrid and the other is from an inbred line.

- *Triple cross hybrids* — result from the crossing of two different three-way cross hybrids.

- *Population hybrids* — result from the crossing of plants or animals in a population with another population. These include crosses between organisms such as interspecific hybrids or crosses between different breeds.

- *Stable hybrid* – a horticultural term which typically refers to an annual plant that, if grown and bred in a small monoculture free of external pollen (e.g., an air-filtered greenhouse) will produce offspring that are "true to type" with respect to phenotype; i.e., a true breeding organism.

- *Hybrid species* – results from hybrid populations evolving reproductive barriers against their parent species through hybrid speciation.

Interspecific Hybrids

Interspecific hybrids are bred by mating two species, normally from within the same genus. The offspring display traits and characteristics of both parents. The offspring of an interspecific cross are very often sterile; thus, hybrid sterility prevents the movement of genes from one species to the other, keeping both species distinct. Sterility is often attributed to the different number of chromosomes the two species have, for example donkeys have 62 chromosomes, while horses have 64 chromosomes, and mules and hinnies have 63 chromosomes. Mules, hinnies, and other normally sterile interspecific hybrids cannot produce viable gametes, because differences in chromosome structure prevent appropriate pairing and segregation during meiosis, meiosis is disrupted, and viable sperm and eggs are not formed. However, fertility in female mules has been reported with a donkey as the father.

Most often other processes occurring in plants and animals keep gametic isolation and species distinction. Species often have different mating or courtship patterns or behaviors, the breeding seasons may be distinct and even if mating does occur antigenic reactions to the sperm of other species prevent fertilization or embryo development. Hybridisation is much more common among organisms that spawn indiscriminately, like soft corals and among plants.

While it is possible to predict the genetic composition of a backcross *on average*, it is not possible to accurately predict the composition of a particular backcrossed individual, due to random segregation of chromosomes. In a species with two pairs of chromosomes, a twice backcrossed individual would be predicted to contain 12.5% of one species' genome (say, species A). However, it may, in fact, still be a 50% hybrid if the chromosomes from species A were lucky in two successive segregations, and meiotic crossovers happened near the telomeres. The chance of this is fairly high: $\left(\dfrac{1}{2}\right)^{(2\times2)} = \dfrac{1}{16}$ (where the "two times two" comes about from two rounds of meiosis with two chromosomes); however, this probability declines markedly with chromosome number and so the actual composition of a hybrid will be increasingly closer to the predicted composition.

Hybrid Species

While not very common, a few animal species have been recognized as being the result of hybridization. The Lonicera fly is an example of a novel animal species that resulted from natural hybridization. The American red wolf appears to be a hybrid species between gray wolf and coyote, although its taxonomic status has been a subject of controversy. The European edible frog appears to be a species, but is actually a semi-permanent hybrid between pool frogs and marsh frogs. The edible frog population is dependent on the presence of at least one of the parents species to be maintained.

Hybrid species of plants are much more common than animals. Many of the crop species are hybrids, and hybridization appears to be an important factor in speciation in some plant groups.

Examples of Hybrid Animals and Animal Populations Derived from Hybrids

Mammals

A "zonkey", a zebra/donkey hybrid.

A "jaglion", a jaguar/lion hybrid.

- Equid hybrids

    o Mule, a cross of female horse and a male donkey.

    o Hinny, a cross between a female donkey and a male horse. Mule and hinny are examples of reciprocal hybrids.

    o Zebroids

        - Zeedonk or Zonkey, a zebra/donkey cross.

        - Zorse, a zebra/horse cross

        - Zony or Zetland, a zebra/pony cross ("zony" is a generic term; "zetland" is specifically a hybrid of the Shetland pony breed with a zebra)

    o hybrid ass, a cross between a donkey and an onager or Asian wild ass.

- Bovid hybrids

    o Dzo, zo or yakow; a cross between a domestic cow/bull and a yak.

    o Beefalo, a cross of an American bison and a domestic cow. This is a fertile breed; this along with genetic evidence has caused them to be recently reclassified into the same genus, *Bos*.

    o Żubroń, a hybrid between wisent (European bison) and domestic cow.

- Sheep-goat hybrid is the cross between a sheep and a goat, which belong to different genera.

- Ursid hybrids, such as the grizzly-polar bear hybrid, occur between black bears, brown bears, and polar bears.

- Felid hybrids

    o Savannah cat are a fertile *breed* developed originally from a cross between the serval [*Leptailurus serval*] and a domestic cat [*Felis catus*].

    o A hybrid between a Bengal tiger and a Siberian tiger is an example of an *intra-specific* hybrid. It also includes the Indochinese tiger, Sumatran tiger too.

    o Pumapards are the hybrid crosses between a puma and a leopard.

    o Ligers and tigons (crosses between a lion and a tiger - the difference in name due to what species the mother and father were - ligers have a lion father and a tiger mother) and other Panthera hybrids such as the lijagulep. Various other wild cat crosses are known involving the lynx, bobcat, leopard, serval, etc.

        - Liligers are the hybrid cross between a male lion and a ligress.

    o Bengals are a fertile *breed* developed originally from a cross between the Asian leopard cat [*Prionailurus bengalensis*] and the domestic cat [*Felis catus*].

- Fertile canid hybrids occur between coyotes, wolves, dingoes, jackals and domestic dogs.

- Hybrids between black and white rhinoceroses have been recognized.

- Hybrid camel, a cross between a bactrian camel and a dromedary camel

- Cama, a cross between a camel and a llama, also an intergeneric hybrid.

- Wholphin, a fertile but very rare cross between a false killer whale and a bottlenose dolphin.

- At Chester Zoo in the United Kingdom, a cross between an African elephant (male) and an Asian elephant (female). The male calf was named Motty. He died of intestinal infection after twelve days.

- Bornean and Sumatran orangutan hybrids have occurred in captivity.

Birds

A mule, a domestic canary/goldfinch hybrid.

- Hybrids between spotted owls and barred owls

- Cagebird breeders sometimes breed hybrids between species of finch, such as goldfinch × canary. These birds are known as mules.

- The perlin is a peregrine falcon – merlin hybrid.

- Gamebird hybrids, hybrids between gamebirds and domestic fowl, including chickens, guineafowl and peafowl, interfamilial hybrids.

- Numerous macaw hybrids and lovebird hybrids are also known in aviculture.

- Red kite × black kite: five bred unintentionally at a falconry center in England. (It is reported that the black kite (the male) refused female black kites but mated with two female red kites.)

- The mulard duck, hybrid of the domestic pekin duck and domesticated muscovy ducks.

- In Australia, New Zealand and other areas where the Pacific black duck occurs, it is hybridised by the much more aggressive introduced mallard. This is a concern to wildlife authorities throughout the affected area, as it is seen as Genetic pollution of the black duck gene pool.

- Hybridisation in gulls is a reasonably frequent occurrence in the wild.

Reptiles

- Hybrid iguana, a single-cross hybrid resulting from natural interbreeding between male marine iguanas and female land iguanas since the late 2000s.

- Crestoua, a cross between a Rhacodactylus Ciliatus (crested gecko) and a Rhacodactylus Chahoua.

- Colubrid snakes of the tribe Lampropeltini have been shown to produce fertile hybrid offspring.

- Hybridization between the endemic Cuban crocodile (*Crocodilus rhombifer*) and the widely distributed American crocodile (*Crocodilus acutus*) is causing conservation problems for the former species as a threat to its genetic integrity.

- Saltwater crocodiles (*Crocodylus porosus*) have mated with Siamese crocodiles (*Crocodylus siamensis*) in captivity producing offspring which in many cases have grown over 20 feet (6.1 metres) in length. It is likely that wild hybridization occurred historically in parts of southeast Asia.

- Many species of boas and pythons are known to produce hybrids, such as carball (a cross between a ball python and a carpet python) or a bloodball (a cross between a blood python and a ball python) however, most of these only occur in captivity. Contrary to popular belief, boa–python hybrids are not possible due to their differing reproductive functions. Boas only produce hybrids with other species of boas, and pythons only produce hybrids with other species of pythons.

Amphibians

- Japanese giant salamanders and Chinese giant salamanders have created hybrids that threaten the survival of Japanese giant salamanders due to the competition for similar resources in Japan.

Fish

- Blood parrot cichlid, which is probably created by crossing a red head cichlid and a Midas cichlid or red devil cichlid

- A group of about 50 hybrids between Australian blacktip shark and the larger common blacktip shark was found by Australia's East Coast in 2012. This is the only known case of hybridization in sharks.

- Silver bream and common bream commonly produce sterile hybrids.

- Tiger muskie is a sterile hybrid between northern pike and muskellunge.

Insects

- Killer bees were created during an attempt to breed a strain of bees that would produce more honey and be better adapted to tropical conditions. This was done by crossing a European honey bee and an African bee.

- The *Colias eurytheme* and *C. philodice* butterflies have retained enough genetic compatibility to produce viable hybrid offspring.

Hybrid Plants

A sterile hybrid between *Trillium cernuum* and *T. grandiflorum*

An ornamental lily hybrid known as *Lilium* 'Citronella'

Many hybrids are created by humans, but natural hybrids occur as well. Plant species hybridize more readily than animal species, and the resulting hybrids are more often fertile hybrids and may reproduce, though there still exist sterile hybrids and selective hybrid elimination where the offspring are less able to survive and are thus eliminated before they can reproduce. A number of plant species are the result of hybridization and polyploidy with many plant species easily cross pollinating and producing viable seeds, the distinction between each species is often maintained by geographical isolation or differences in the flowering period. Since plants hybridize frequently without much work, they are often created by humans in order to produce improved plants. These improvements can include the production of more or improved seeds, fruits or other plant parts for consumption, or to make a plant more winter or heat hardy or improve its growth and/or appearance for use in horticulture. Much work is now being done with hybrids to produce more disease resistant plants for both agricultural and horticultural crops. In many groups of plants hybridization has been used to produce larger and more showy flowers and new flower colors. Hybridization may be restricted to the desired parent species through the use of pollination bags.

Many plant genera and species have their origins in polyploidy. Autopolyploidy results from the sudden multiplication in the number of chromosomes in typical normal populations caused by unsuccessful separation of the chromosomes during meiosis. Tetraploids (plants with four sets of chromosomes rather than two) are common in a number of different groups of plants and over time these plants can differentiate into distinct species from the normal diploid line. In *Oenothera lamarchiana* the diploid species has 14 chromosomes, this species has spontaneously given rise to plants with 28 chromosomes that have been given the name *Oenothera gigas*. When hybrids are formed between the tetraploids and the diploid population, the resulting offspring tend to be sterile triploids, thus effectively stopping the intermixing of genes between the two groups of plants (unless the diploids, in rare cases, produce unreduced gametes).

Another form of polyploidy called allopolyploidy occurs when two different species mate and produce polyploid hybrids. Usually the typical chromosome number is doubled, and the four sets of chromosomes can pair up during meiosis, thus the polyploids can produce offspring. Usually, these offspring can mate and reproduce with each other but cannot back-cross with the parent species. Allopolyploids may be able to adapt to new habitats that neither of their parent species inhabited

Heterosis

Hybrids are sometimes stronger than either parent variety, a phenomenon most common with plant hybrids, which when present is known as *hybrid vigor* (heterosis) or heterozygote advantage. A transgressive phenotype is a phenotype displaying more extreme characteristics than either of the parent lines. Plant breeders make use of a number of techniques to produce hybrids, including line breeding and the formation of complex hybrids. An economically important example is hybrid maize (corn), which provides a considerable seed yield advantage over open pollinated varieties. Hybrid seed dominates the commercial maize seed market in the United States, Canada and many other major maize producing countries.

Examples of Plant Hybrids

The multiplication symbol × (not italicised) indicates a hybrid in the Latin binomial nomenclature. Placed before the binomial it indicates a hybrid between species from different genera (intergeneric hybrid):-

- × *Fatshedera lizei*, a hybrid between *Hedera helix* and *Fatsia japonica*

- × *Heucherella*, a hybrid genus between *Heuchera* and *Tiarella*

- × *Philageria veitchii* is a hybrid between *Lapageria rosea* and *Philesia magellanica*; it is more similar in appearance to the former

- Leyland cypress, [× *Cupressocyparis leylandii*] hybrid between Monterey cypress and Nootka cypress

- Triticale, [× *Triticosecale*] a wheat–rye hybrid

- × *Urceocharis*, a hybrid between *Eucharis* and *Urceolina*

Interspecific plant hybrids include:

- *Dianthus* × *allwoodii* (*Dianthus caryophyllus* × *Dianthus plumarius*)

- Limequat *Citrus* × *floridana*, key lime *Citrus aurantiifolia* and kumquat *Citrus japonica* hybrid

- Loganberry *Rubus* × *loganobaccus*, a hybrid between raspberry *Rubus idaeus* and blackberry *Rubus ursinus*

- London plane (*Platanus orientalis* × *Platanus occidentalis*), thus forming *Platanus* × *acerifolia*

- *Magnolia* × *alba* (*Magnolia champaca* × *Magnolia montana*)

- Peppermint, a hybrid between spearmint and water mint

- *Quercus* × *warei* (*Quercus robur* × *Quercus bicolor*) 'Nadler' (marketed in the United States under the trade name Kindred Spirit hybrid oak)

- Tangelo, a hybrid of a Mandarin orange and a pomelo which may have been developed in Asia about 3,500 years ago

- Wheat; most modern and ancient wheat breeds are themselves hybrids. Bread wheat is a hexaploid hybrid of three wild grasses; durum (pasta) wheat is a tetraploid hybrid of two wild grasses

- Grapefruit, hybrid between a pomelo and the Jamaican sweet orange

Some natural hybrids:

- *Iris albicans*, a sterile hybrid which spreads by rhizome division

- Evening primrose, a flower which was the subject of famous experiments by Hugo de Vries on polyploidy and diploidy

Hybrids in Nature

Hybridization between two closely related species is actually a common occurrence in nature but is also being greatly influenced by anthropogenic changes as well. Hybridization is a naturally occurring genetic process where individuals from two genetically distinct populations mate. As stated above, it can occur both intraspecifically, between different distinct populations within the same species, and interspecifically, between two different species. Hybrids can be either sterile/ not viable or viable/fertile. This affects the kind of effect that this hybrid will have on its and other populations that it interacts with. Many hybrid zones are known where the ranges of two species meet, and hybrids are continually produced in great numbers. These hybrid zones are useful as biological model systems for studying the mechanisms of speciation (Hybrid speciation). Recently DNA analysis of a bear shot by a hunter in the North West Territories confirmed the existence of naturally-occurring and fertile grizzly–polar bear hybrids.

Anthropogenic Hybridization

Changes to the environment caused by humans, such as fragmentation and Introduced species, are becoming more widespread. This increases the challenges in managing certain populations that are experiencing introgression, and is a focus of conservation genetics.

Introduced Species and Habitat Fragmentation

Humans have introduced species worldwide to environments for a long time, both intentionally such as establishing a population to be used as a biological control, and unintentionally such as accidental escapes of individuals out of agriculture. This causes drastic global effects on various populations, including through hybridization.

When habitats become broken apart, one of two things can occur, genetically speaking. The first is that populations that were once connected can be cut off from one another, preventing their genes from interacting. Occasionally, this will result in a population of one species breeding with a population of another species as a means of surviving such as the case with the red wolves. Their population numbers being so small, they needed another means of survival. Habitat fragmentation also led to the influx of generalist species into areas where they would not have been, leading to competition and in some cases interbreeding/incorporation of a population into another. In this way, habitat fragmentation is essentially an indirect method of introducing species to an area.

The Hybridization Continuum

There is a kind of continuum with three semi-distinct categories dealing with anthropogenic hybridization: hybridization without Introgression, hybridization with widespread introgression, and essentially a Hybrid swarm. Depending on where a population falls along this continuum, the management plans for that population will change. Hybridization is currently an area of great discussion within Wildlife management and habitat management. Global climate change is creating other changes such as difference in population distributions which are indirect causes for an increase in anthropogenic hybridization.

Consequences

Hybridization can be a less discussed way toward extinction than within detection of where a population lies along the hybrid continuum. The dispute of hybridization is how to manage the resulting hybrids. When a population experiences hybridization with substantial introgression, there still exists parent types of each set of individuals. When a complete hybrid swarm is created, all the individuals are hybrids.

Management of Hybrids

Conservationists disagree on when is the proper time to give up on a population that is becoming a hybrid swarm or to try and save the still existing pure individuals. Once it becomes a complete mixture, we should look to conserve those hybrids to avoid their loss. Most leave it as a case-by-case basis, depending on detecting of hybrids within the group. It is nearly impossible to regulate hybridization via policy because hybridization can occur beneficially when it occurs "naturally" and there is the matter of protecting those previously mentioned hybrid swarms because if they are the only remaining evidence of prior species, they need to be conserved as well.

Expression of Parental Traits in Hybrids

When two distinct types of organisms breed with each other, the resulting hybrids typically have intermediate traits (e.g., one parent has red flowers, the other has white, and the hybrid, pink flowers). Commonly, hybrids also combine traits seen only separately in one parent or the other (e.g., a bird hybrid might combine the yellow head of one parent with the orange belly of the other).

In a hybrid, any trait that falls outside the range of parental variation is termed heterotic. Heterotic hybrids do have new traits, that is, they are not intermediate. *Positive heterosis* produces more robust hybrids, they might be stronger or bigger; while the term *negative heterosis* refers to weaker or smaller hybrids. Heterosis is common in both animal and plant hybrids. For example, hybrids between a lion and a tigress ("ligers") are much larger than either of the two progenitors, while a tigon (lioness × tiger) is smaller. Also the hybrids between the common pheasant (*Phasianus colchicus*) and domestic fowl (*Gallus gallus*) are larger than either of their parents, as are those produced between the common pheasant and hen golden pheasant (*Chrysolophus pictus*). Spurs are absent in hybrids of the former type, although present in both parents.

Genetic Mixing and Extinction

Regionally developed ecotypes can be threatened with extinction when new alleles or genes are introduced that alter that ecotype. This is sometimes called genetic mixing. Hybridization and introgression of new genetic material can lead to the replacement of local genotypes if the hybrids are more fit and have breeding advantages over the indigenous ecotype or species. These hybridization events can result from the introduction of non native genotypes by humans or through habitat modification, bringing previously isolated species into contact. Genetic mixing can be especially detrimental for rare species in isolated habitats, ultimately affecting the population to such a degree that none of the originally genetically distinct population remains.

Effect on Biodiversity and Food Security

In agriculture and animal husbandry, the Green Revolution's use of conventional hybridization increased yields by breeding "high-yielding varieties". The replacement of locally indigenous breeds, compounded with unintentional cross-pollination and crossbreeding (genetic mixing), has reduced the gene pools of various wild and indigenous breeds resulting in the loss of genetic diversity. Since the indigenous breeds are often well-adapted to local extremes in climate and have immunity to local pathogens this can be a significant genetic erosion of the gene pool for future breeding. Therefore, commercial plant geneticists strive to breed "widely adapted" cultivars to counteract this tendency.

Limiting Factors

A number of conditions exist that limit the success of hybridization, the most obvious is great genetic diversity between most species. But in animals and plants that are more closely related hybridization barriers can include morphological differences, differing times of fertility, mating behaviors and cues, physiological rejection of sperm cells or the developing embryo.

In plants, barriers to hybridization include blooming period differences, different pollinator vectors, inhibition of pollen tube growth, somatoplastic sterility, cytoplasmic-genic male sterility and structural differences of the chromosomes.

Mythical, Legendary and Religious Hybrids

Ancient folktales often contain mythological creatures, sometimes these are described as hybrids (e.g., hippogriff as the offspring of a griffin and a horse, and the Minotaur which is the offspring of Pasiphaë and a white bull). More often they are kind of chimera, i.e., a composite of the physical attributes of two or more kinds of animals, mythical beasts, and often humans, with no suggestion that they are the result of interbreeding, e.g., harpies, mermaids, and centaurs.

In the Bible, the Old Testament contains several passages which talk about a first generation of hybrid giants who were known as the Nephilim. The Book of Genesis (6:4) states that "the sons of God went to the daughters of humans and had children by them". As a result, the offspring was born as hybrid giants who became mighty heroes of old and legendary famous figures of ancient times. In addition, the Book of Numbers (13:33) says that the descendants of Anak came from the Nephilim, whose bodies looked exactly like men, but with an enormous height. According to the apocryphal Book of Enoch the Nephilim were wicked sons of fallen angels who had lusted with attractive women.

Biosynthesis

Biosynthesis (also called biogenesis or anabolism) is a multi-step, enzyme-catalyzed process where substrates are converted into more complex products in living organisms. In biosynthesis, simple compounds are modified, converted into other compounds, or joined together to form macromolecules. This process often consists of metabolic pathways. Some of these biosynthetic pathways are located within a single cellular organelle, while others involve enzymes that are located within multiple cellular organelles. Examples of these biosynthetic pathways include the production of lipid membrane components and nucleotides.

The prerequisite elements for biosynthesis include: precursor compounds, chemical energy (e.g. ATP), and catalytic enzymes which may require coenzymes (e.g.NADH, NADPH). These elements create monomers, the building blocks for macromolecules. Some important biological macromolecules include: proteins, which are composed of amino acid monomers joined via peptide bonds, and DNA molecules, which are composed of nucleotides joined via phosphodiester bonds.

Properties of Chemical Reactions

Biosynthesis occurs due to a series of chemical reactions. For these reactions to take place, the following elements are necessary:

- Precursor compounds: these compounds are the starting molecules or substrates in a reaction. These may also be viewed as the reactants in a given chemical process.

- Chemical energy: chemical energy can be found in the form of high energy molecules. These molecules are required for energetically unfavorable reactions. Furthermore, the hydrolysis of these compounds drives a reaction forward. High energy molecules, such as ATP, have three phosphates. Often, the terminal phosphate is split off during hydrolysis and transferred to another molecule.

- Catalytic enzymes: these molecules are special proteins that catalyze a reaction by increasing the rate of the reaction and lowering the activation energy.

- Coenzymes or cofactors: cofactors are molecules that assist in chemical reactions. These may be metal ions, vitamin derivatives such as NADH and acetyl CoA, or non-vitamin derivatives such as ATP. In the case of NADH, the molecule transfers a hydrogen, whereas acetyl CoA transfers an acetyl group, and ATP transfers a phosphate.

In the simplest sense, the reactions that occur in biosynthesis have the following format:

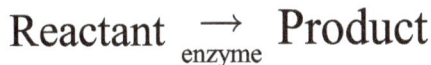

$$\text{Reactant} \xrightarrow[\text{enzyme}]{} \text{Product}$$

Some variations of this basic equation which will be discussed later in more detail are:

1. Simple compounds which are converted into other compounds, usually as part of a multiple step reaction pathway. Two examples of this type of reaction occur during the formation of nucleic acids and the charging of tRNA prior to translation. For some of these steps, chemical energy is required:

$$\text{Precursor molecule} + \text{ATP} \rightleftharpoons \text{product AMP} + \text{PP}_i$$

2. Simple compounds that are converted into other compounds with the assistance of cofactors. For example, the synthesis of phospholipids requires acetyl CoA, while the synthesis of another membrane component, shingolipids, requires NADH and FADH for the formation the sphingosine backbone. The general equation for these examples is:

$$\text{Precursor molecule} + \text{Cofactor} \xrightarrow[\text{enzyme}]{} \text{macromolecule}$$

3. Simple compounds that join together to create a macromolecule. For example, fatty acids join together to form phopspholipids. In turn, phospholipids and cholesterol interact non-covalently in order to form the lipid bilayer. This reaction may be depicted as follows:

$$\text{Molecule 1} + \text{Molecule 2} \rightarrow \text{macromolecule}$$

Lipids

Individual Lipid

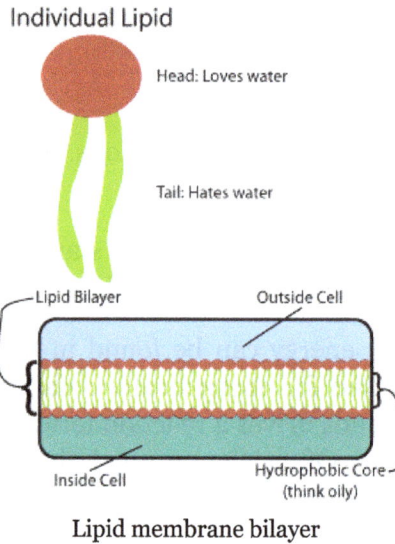

Head: Loves water

Tail: Hates water

Lipid Bilayer   Outside Cell

Inside Cell   Hydrophobic Core
                     (think oily)

Lipid membrane bilayer

Many intricate macromolecules are synthesized in a pattern of simple, repeated structures. For example, the simplest structures of lipids are fatty acids. Fatty acids are hydrocarbon derivatives; they contain a carboxyl group "head" and a hydrocarbon chain "tail." These fatty acids create larger components, which in turn incorporate noncovalent interactions to form the lipid bilayer. Fatty acid chains are found in two major components of membrane lipids: phospholipids and sphingolipids. A third major membrane component, cholesterol, does not contain these fatty acid units.

Phospholipids

The foundation of all biomembranes consists of a bilayer structure of phospholipids. The phospholipid molecule is amphipathic; it contains a hydrophilic polar head and a hydrophobic nonpolar tail. The phospholipid heads interact with each other and aqueous media, while the hydrocarbon tails orient themselves in the center, away from water. These latter interactions drive the bilayer structure that acts as a barrier for ions and molecules.

There are various types of phospholipids; consequently, their synthesis pathways differ. However, the first step in phospholipid synthesis involves the formation of phosphatidate or diacylglycerol 3-phosphate at the endoplasmic reticulum and outer mitochondrial membrane. The synthesis pathway is found below:

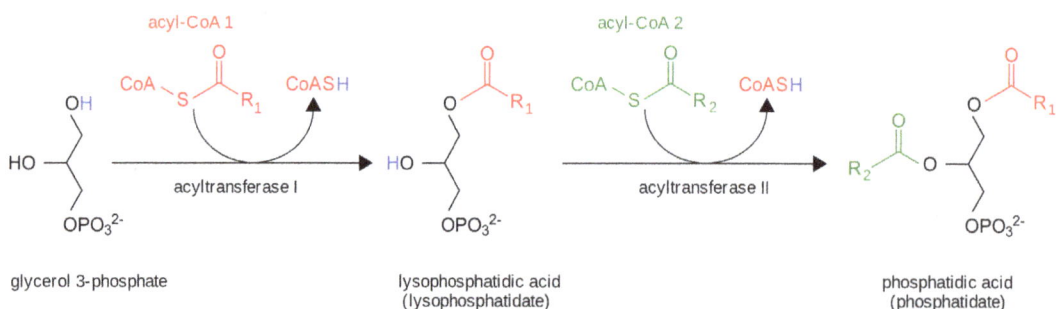

glycerol 3-phosphate

lysophosphatidic acid
(lysophosphatidate)

phosphatidic acid
(phosphatidate)

The pathway starts with glycerol 3-phosphate, which gets converted to lysophosphatidate via the addition of a fatty acid chain provided by acyl coenzyme A. Then, lysophosphatidate is converted to phosphatidate via the addition of another fatty acid chain contributed by a second acyl CoA; all of these steps are catalyzed by the glycerol phosphate acyltransferase enzyme. Phospholipid synthesis continues in the endoplasmic reticulum, and the biosynthesis pathway diverges depending on the components of the particular phospholipid.

Sphingolipids

Like phospholipids, these fatty acid derivatives have a polar head and nonpolar tails. Unlike phospholipids, sphingolipids have a sphingosine backbone. Sphingolipids exist in eukaryotic cells and are particularly abundant in the central nervous system. For example, sphingomyelin is part of the myelin sheath of nerve fibers.

Sphingolipids are formed from ceramides that consist of a fatty acid chain attached to the amino group of a sphingosine backbone. These ceramides are synthesized from the acylation of sphingosine. The biosynthetic pathway for sphingosine is found below:

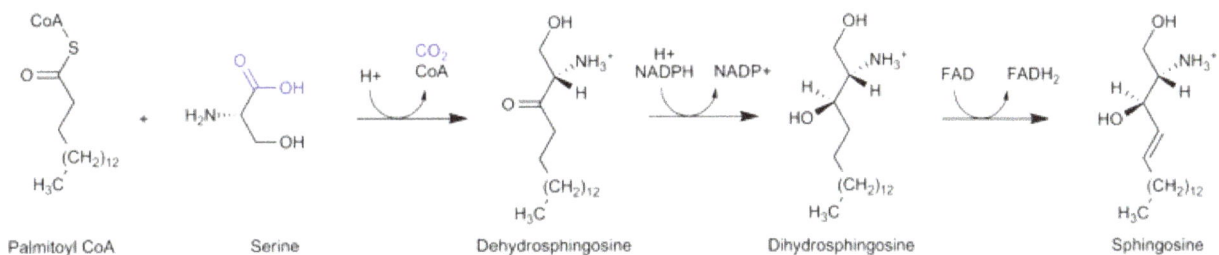

| Palmitoyl CoA | Serine | Dehydrosphingosine | Dihydrosphingosine | Sphingosine |

As the image denotes, during sphingosine synthesis, palmitoyl CoA and serine undergo a condensation reaction which results in the formation of dehydrosphingosine. This product is then reduced to form dihydrospingosine, which is converted to sphingosine via the oxidation reaction by FAD.

Cholesterol

This lipid belongs to a class of molecules called sterols. Sterols have four fused rings and a hydroxyl group. Cholesterol is a particularly important molecule. Not only does it serve as a component of lipid membranes, it is also a precursor to several steroid hormones, including cortisol, testosterone, and estrogen.

More generally, this synthesis occurs in three stages, with the first stage taking place in the cytoplasm and the second and third stages occurring in the endoplasmic reticulum. The stages are as follows:

1. The synthesis of isopentenyl pyrophosphate, the "building block" of cholesterol

2. The formation of squalene via the condensation of six molecules of isopentenyl phosphate

3. The conversion of squalene into cholesterol via several enzymatic reactions

Cholesterol is synthesized from acetyl CoA. The pathway is shown below:

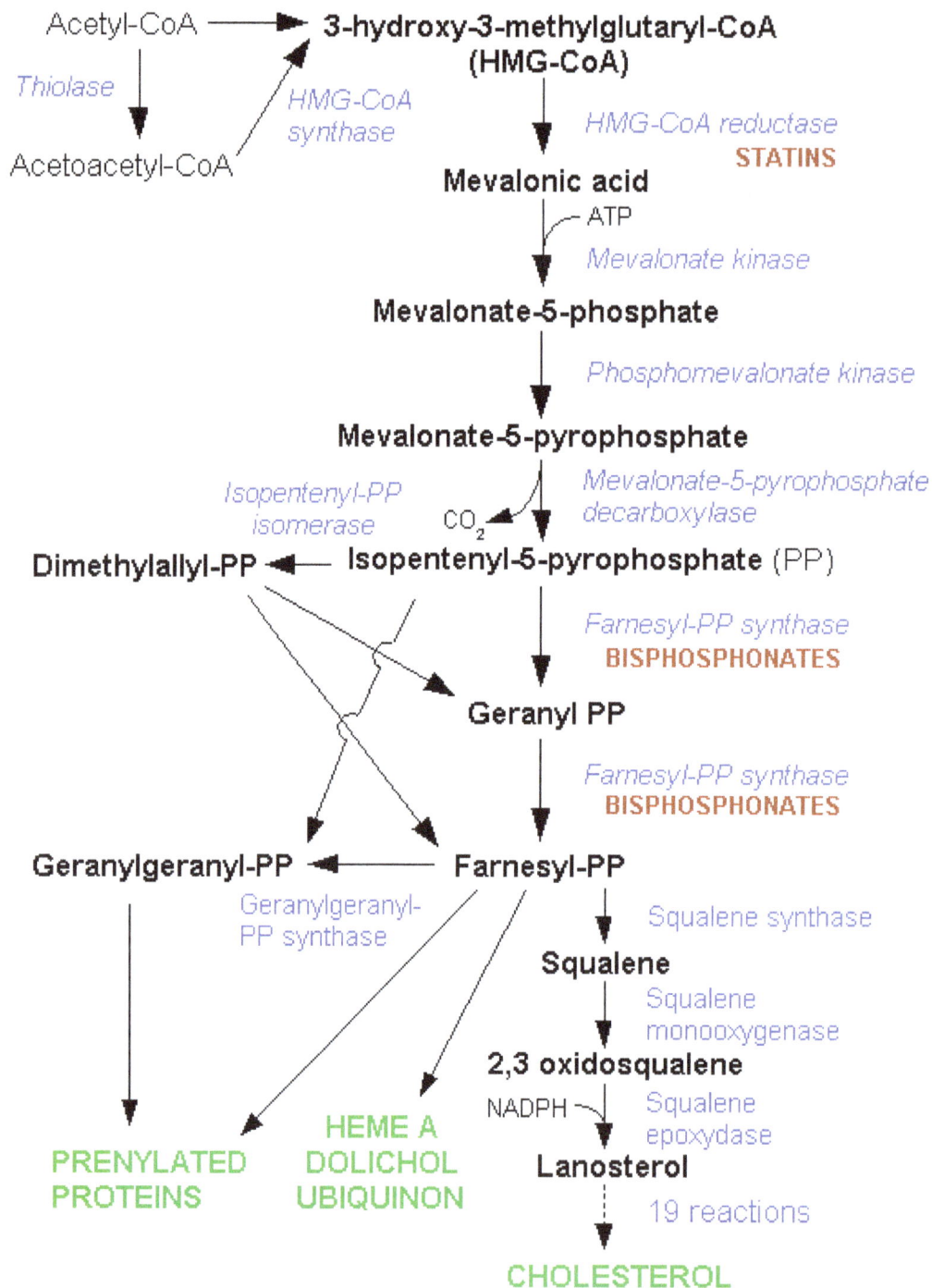

Acetyl-CoA ⟶ **3-hydroxy-3-methylglutaryl-CoA (HMG-CoA)**

Thiolase

HMG-CoA synthase

Acetoacetyl-CoA

HMG-CoA reductase
STATINS

Mevalonic acid

ATP

Mevalonate kinase

Mevalonate-5-phosphate

Phosphomevalonate kinase

Mevalonate-5-pyrophosphate

Isopentenyl-PP isomerase

CO_2

Mevalonate-5-pyrophosphate decarboxylase

Dimethylallyl-PP ⟵ **Isopentenyl-5-pyrophosphate** (PP)

Farnesyl-PP synthase
BISPHOSPHONATES

Geranyl PP

Farnesyl-PP synthase
BISPHOSPHONATES

Geranylgeranyl-PP ⟵ **Farnesyl-PP**

Geranylgeranyl-PP synthase

Squalene synthase

Squalene

Squalene monooxygenase

2,3 oxidosqualene

NADPH

Squalene epoxydase

Lanosterol

PRENYLATED PROTEINS

HEME A DOLICHOL UBIQUINON

19 reactions

CHOLESTEROL

Nucleotides

The biosynthesis of nucleotides involves enzyme-catalyzed reactions that convert substrates into more complex products. Nucleotides are the building blocks of DNA and RNA. Nucleotides are composed of a five-membered ring formed from ribose sugar in RNA, and deoxyribose sugar in DNA; these sugars are linked to a purine or pyrimidine base with a glycosidic bond and a phosphate group at the 5' location of the sugar.

Purine Nucleotides

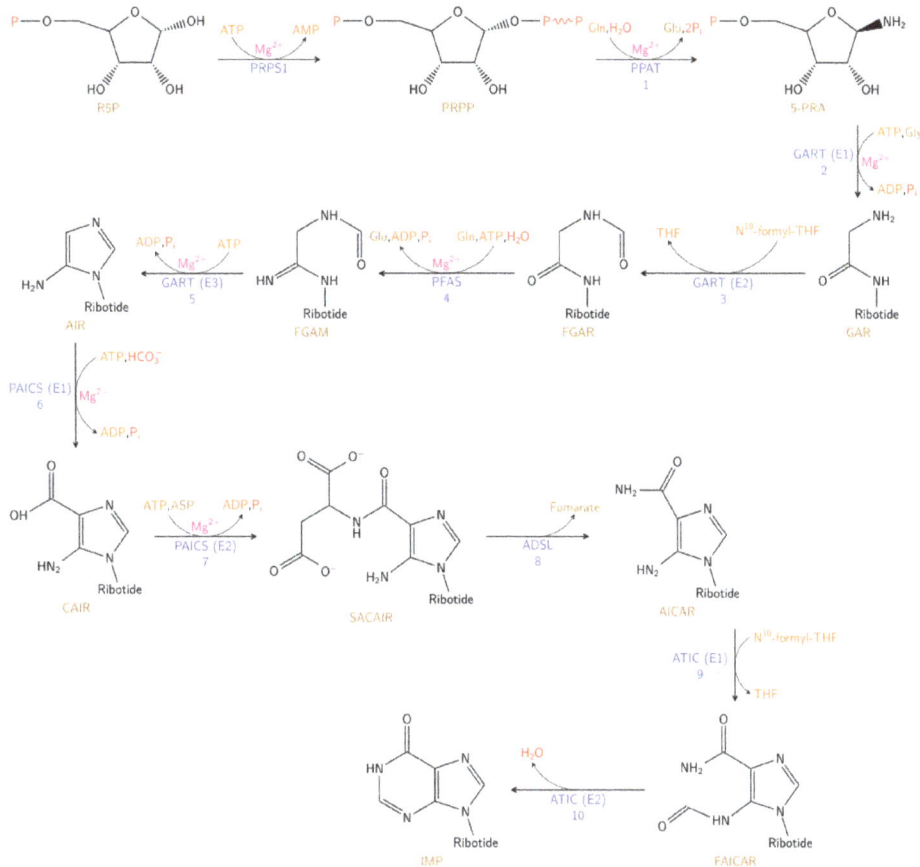

The Synthesis of IMP.

The DNA nucleosides adenosine and guanosine consist of a purine base attached to a ribose sugar with a glycosidic bond. In the case of RNA nucleotides deoxyadenosine and deoxyguanosine, the purine bases are attached to a deoxyribose sugar with a glycosidic bond. The purine bases on DNA and RNA nucleotides are synthesized in a twelve-step reaction mechanism present in most single-celled organisms. Higher eukaryotes employ a similar reaction mechanism in ten reaction steps. Purine bases are synthesized by converting phosphoribosyl pyrophosphate (PRPP) to inosine monophosphate (IMP), which is the first key intermediate in purine base biosynthesis. Further enzymatic modification of IMP produces the adenosine and guanosine bases of nucleotides.

1. The first step in purine biosynthesis is a condensation reaction, performed by glutamine-PRPP amidotransferase. This enzyme transfers the amino group from glutamine to PRPP, forming 5-phosphoribosylamine. The following step requires the activation of glycine by the addition of a phosphate group from ATP.

2. GAR synthetase performs the condensation of activated glycine onto PRPP, forming glycineamide ribonucleotide (GAR).

3. GAR transformylase adds a formyl group onto the amino group of GAR, forming formylglycinamide ribonucleotide (FGAR).

4. FGAR amidotransferase catalyzes the addition of a nitrogen group to FGAR, forming formylglycinamidine ribonucleotide (FGAM).

5. FGAM cyclase catalyzes ring closure, which involves removal of a water molecule, forming the 5-membered imidazole ring 5-aminoimidazole ribonucleotide (AIR).

6. N5-CAIR synthetase transfers a carboxyl group, forming the intermediate N5-carboxyaminoimidazole ribonucleotide (N5-CAIR).

7. N5-CAIR mutase rearranges the carboxyl functional group and transfers it onto the imidazole ring, forming carboxyamino- imidazole ribonucleotide (CAIR). The two step mechanism of CAIR formation from AIR is mostly found in single celled organisms. Higher eukaryotes contain the enzyme AIR carboxylase, which transfers a carboxyl group directly to AIR imidazole ring, forming CAIR.

8. SAICAR synthetase forms a peptide bond between aspartate and the added carboxyl group of the imidazole ring, forming N-succinyl-5-aminoimidazole-4-carboxamide ribonucleotide (SAICAR).

9. SAICAR lyase removes the carbon skeleton of the added aspartate, leaving the amino group and forming 5-aminoimidazole-4-carboxamide ribonucleotide (AICAR).

10. AICAR transformylase transfers a carbonyl group to AICAR, forming N-formylaminoimidazole- 4-carboxamide ribonucleotide (FAICAR).

11. The final step involves the enzyme IMP synthase, which performs the purine ring closure and forms the inosine monophosphate (IMP) intermediate.

Pyrimidine Nucleotides

Other DNA and RNA nucleotide bases that are linked to the ribose sugar via a glycosidic bond are thymine, cytosine and uracil (which is only found in RNA). Uridine monophosphate biosynthesis involves an enzyme that is located in the mitochondrial inner membrane and multifunctional enzymes that are located in the cytosol.

1. The first step involves the enzyme carbamoyl phosphate synthase combining glutamine with CO_2 in an ATP dependent reaction to form carbamoyl phosphate.

2. Aspartate carbamoyltransferase condenses carbamoyl phosphate with aspartate to form uridosuccinate.

3. Dihydroorotase performs ring closure, a reaction that loses water, to form dihydroorotate.

4. Dihydroorotate dehydrogenase, located within the mitochondrial inner membrane, oxidizes dihydroorotate to orotate.

5. Orotate phosphoribosyl hydrolase (OMP pyrophosphorylase) condenses orotate with PRPP to form orotidine-5'-phosphate.

6. OMP decarboxylase catalyzes the conversion of orotidine-5'-phosphate to UMP.

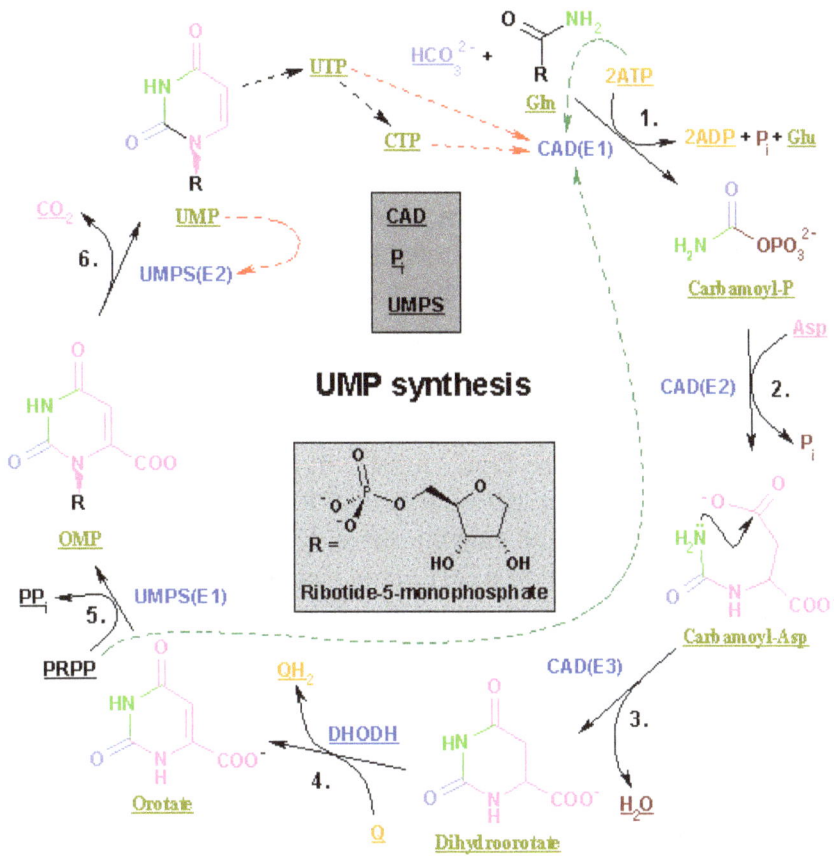

Uridine monophosphate (UMP) biosynthesis

After the uridine nucleotide base is synthesized, the other bases, cytosine and thymine are synthesized. Cytosine biosynthesis is a two-step reaction which involves the conversion of UMP to UTP. Phosphate addition to UMP is catalyzed by a kinase enzyme. The enzyme CTP synthase catalyzes the next reaction step: the conversion of UTP to CTP by transferring an amino group from glutamine to uridine; this forms the cytosine base of CTP. The mechanism, which depicts the reaction UTP + ATP + glutamine ⇔ CTP + ADP + glutamate, is below:

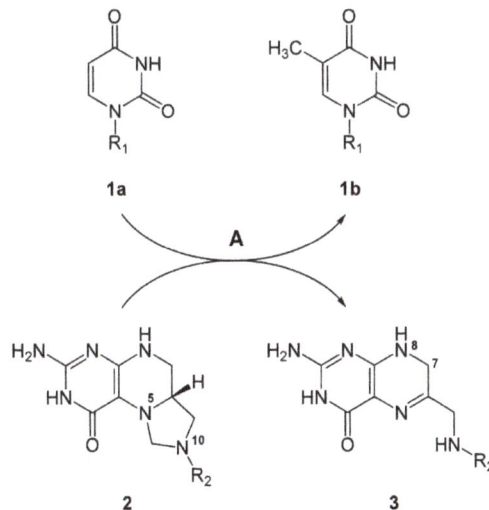

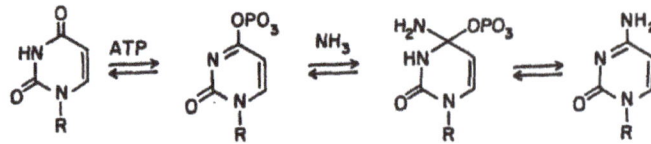

Cytosine is a nucleotide that is present in both DNA and RNA. However, uracil is only found in RNA. Therefore, after UTP is synthesized, it is must be converted into a deoxy form to be incorporated into DNA. This conversion involves the enzyme ribonucleoside triphosphate reductase. This reaction that removes the 2'-OH of the ribose sugar to generate deoxyribose is not affected by the bases attached to the sugar. This non-specificity allows ribonucleoside triphosphate reductase to convert all nucleotide triphosphates to deoxyribonucleotide by a similar mechanism.

In contrast to uracil, thymine bases are found mostly in DNA, not RNA. Cells do not normally contain thymine bases that are linked to ribose sugars in RNA, thus indicating that cells only synthesize deoxyribose-linked thymine. The enzyme thymidylate synthetase is responsible for synthesizing thymine residues from dUMP to dTMP. This reaction transfers a methyl group onto the uracil base of dUMP to generate dTMP. The thymidylate synthase reaction, dUMP + 5,10-methylenetetrahydrofolate ⇔ dTMP + dihydrofolate, is shown to the right.

DNA

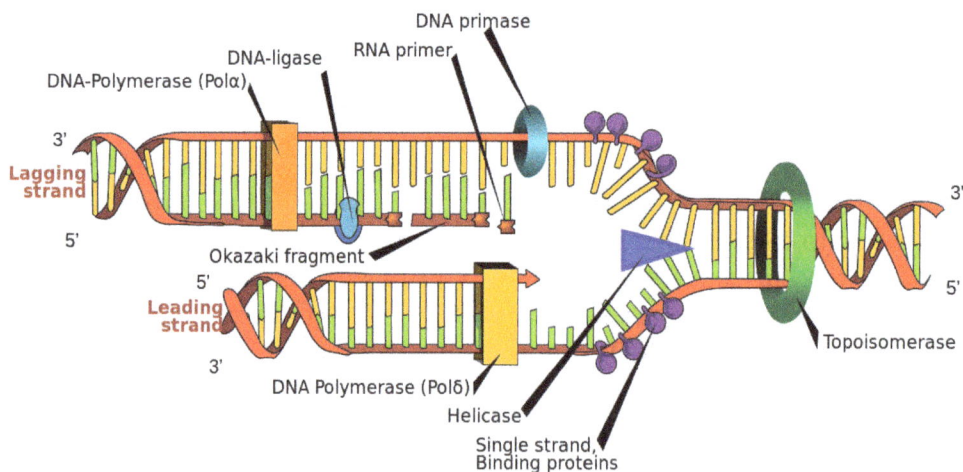

As DNA polymerase moves in a 3' to 5' direction along the template strand, it synthesizes a new strand in the 5' to 3' direction

Although there are differences between eukaryotic and prokaryotic DNA synthesis, the following section denotes key characteristics of DNA replication shared by both organisms.

DNA is composed of nucleotides that are joined by phosphodiester bonds. DNA synthesis, which takes place in the nucleus, is a semiconservative process, which means that the resulting DNA molecule contains an original strand from the parent structure and a new strand. DNA synthesis is catalyzed by a family of DNA polymerases that require four deoxynucleoside triphosphates, a template strand, and a primer with a free 3'OH in which to incorporate nucleotides.

In order for DNA replication to occur, a replication fork is created by enzymes called helicases which unwind the DNA helix. Topoisomerases at the replication fork remove supercoils caused by DNA unwinding, and single-stranded DNA binding proteins maintain the two single-stranded DNA templates stabilized prior to replication.

DNA synthesis is initiated by the RNA polymerase primase, which makes an RNA primer with a free 3'OH. This primer is attached to the single-stranded DNA template, and DNA polymerase elongates the chain by incorporating nucleotides; DNA polymerase also proofreads the newly synthesized DNA strand.

During the polymerization reaction catalyzed by DNA polymerase, a nucleophilic attack occurs by the 3'OH of the growing chain on the innermost phosphorus atom of a deoxynucleoside triphosphate; this yields the formation of a phosphodiester bridge that attaches a new nucleotide and releases pyrophosphate.

Two types of strands are created simultaneously during replication: the leading strand, which is synthesized continuously and grows towards the replication fork, and the lagging strand, which is made discontinuously in Okazaki fragments and grows away from the replication fork. Okazaki fragments are covalently joined by DNA ligase to form a continuous strand. Then, to complete DNA replication, RNA primers are removed, and the resulting gaps are replaced with DNA and joined via DNA ligase.

Amino Acids

A protein is a polymer that is composed from amino acids that are linked by peptide bonds. There are more than 300 amino acids found in nature of which only twenty, known as the standard amino acids, are the building blocks for protein. Only green plants and most microbes are able to synthesize all of the 20 standard amino acids that are needed by all living species. Mammals can only synthesize ten of the twenty standard amino acids. The other amino acids, valine, methionine, leucine, isoleucine, phenylalanine, lysine, threonine and tryptophan for adults and histidine, and arginine for babies are obtained through diet.

Amino Acid Basic Structure

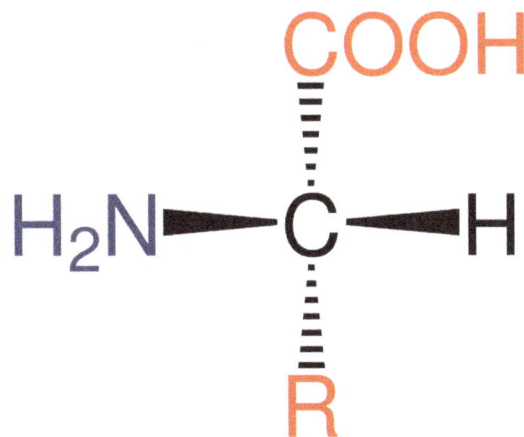

$$H_2N-C-H$$

with COOH above the central C and R below it.

L-amino acid

The general structure of the standard amino acids includes a primary amino group, a carboxyl group and the functional group attached to the α-carbon. The different amino acids are identified by the functional group. As a result of the three different groups attached to the α-carbon, amino acids are asymmetrical molecules. For all standard amino acids, except glycine, the α-carbon is a chiral center. In the case of glycine, the α-carbon has two hydrogen atoms, thus adding symmetry to this molecule. With the exception of proline, all of the amino acids found in life have the L-iso-form conformation. Proline has a functional group on the α-carbon that forms a ring with the amino group.

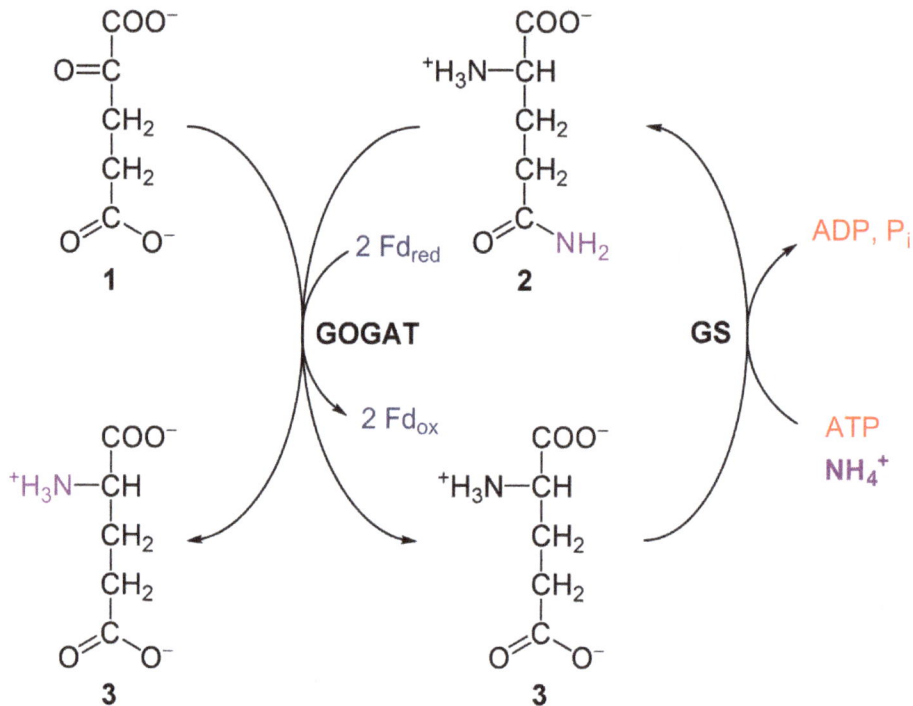

Nitrogen Source

One major step in amino acid biosynthesis involves incorporating a nitrogen group onto the α-carbon. In cells, there are two major pathways of incorporating nitrogen groups. One pathway involves the enzyme glutamine oxoglutarate aminotransferase (GOGAT) which removes the amide amino group of glutamine and transfers it onto 2-oxoglutarate, producing two glutamate molecules. In this catalysis reaction, glutamine serves as the nitrogen source. An image illustrating this reaction is found to the right.

The other pathway for incorporating nitrogen onto the α-carbon of amino acids involves the enzyme glutamate dehydrogenase (GDH). GDH is able to transfer ammonia onto 2-oxoglutarate and form glutamate. Furthermore, the enzyme glutamine synthetase (GS) is able to transfer ammonia onto glutamate and synthesize glutamine, replenishing glutamine.

The Glutamate Family of Amino Acids

The glutamate family of amino acids includes the amino acids that derive from the amino acid glutamate. This family includes: glutamate, glutamine, proline, and arginine. This family also includes the amino acid lysine, which is derived from α-ketoglutarate.

The biosynthesis of glutamate and glutamine is a key step in the nitrogen assimilation discussed above. The enzymes GOGAT and GDH catalyze the nitrogen assimilation reactions.

In bacteria, the enzyme glutamate 5-kinase initiates the biosynthesis of proline by transferring a phosphate group from ATP onto glutamate. The next reaction is catalyzed by the enzyme pyrroline-5-carboxylate synthase (P5CS), which catalyzes the reduction of the γ-carboxyl group of L-glutamate 5-phosphate. This results in the formation of glutamate semialdehyde, which spontaneously cyclizes to pyrroline-5-carboxylate. Pyrroline-5-carboxylate is further reduced by the enzyme pyrroline-5-carboxylate reductase (P5CR) to yield a proline amino acid.

In the first step of arginine biosynthesis in bacteria, glutamate is acetylated by transferring the acetyl group from acetyl-CoA at the N-α position; this prevents spontaneous cyclization. The enzyme N-acetylglutamate synthase (glutamate N-acetyltransferase) is responsible for catalyzing the acetylation step. Subsequent steps are catalyzed by the enzymes N-acetylglutamate kinase, N-acetyl-gamma-glutamyl-phosphate reductase, and acetylornithine/succinyldiamino pimelate aminotransferase and yield the N-acetyl-L-ornithine. The acetyl group of acetylornithine is removed by the enzyme acetylornithinase (AO) or ornithine acetyltransferase (OAT), and this yields ornithine. Then, the enzymes citrulline and argininosuccinate convert ornithine to arginine.

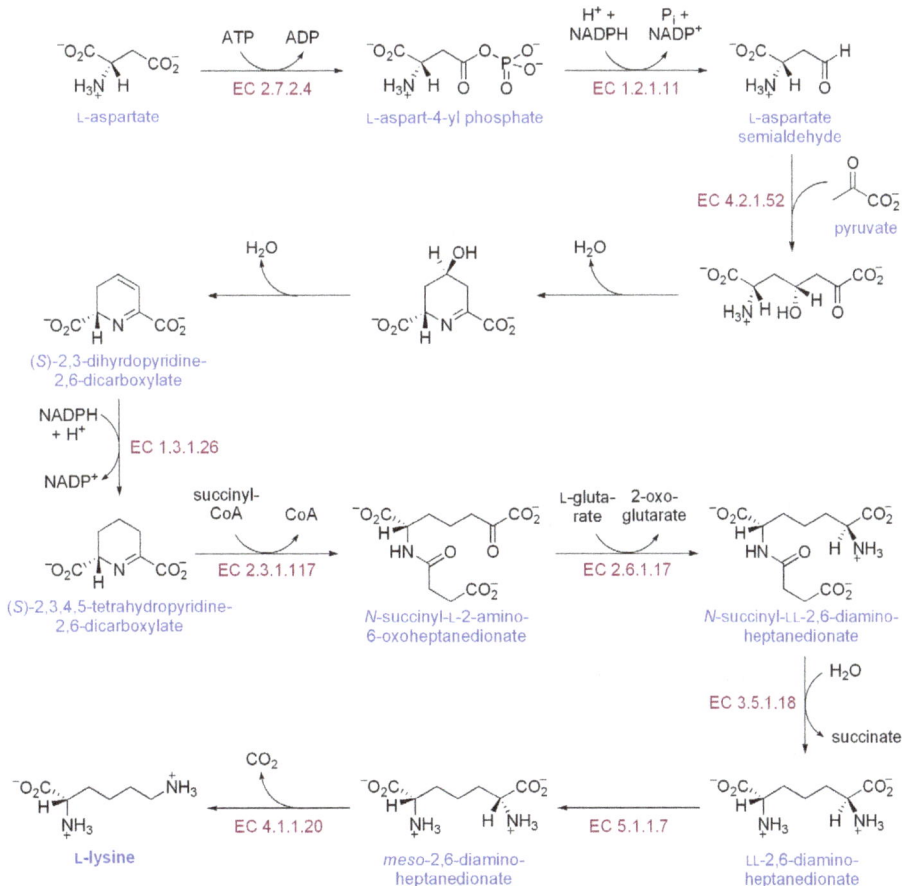

The diaminopimelic acid pathway

There are two distinct lysine biosynthetic pathways: the diaminopimelic acid pathway and the α-aminoadipate pathway. The most common of the two synthetic pathways is the diaminopimelic acid pathway; it consists of several enzymatic reactions that add carbon groups to aspartate to yield lysine:

1. Aspartate kinase initiates the diaminopimelic acid pathway by phosphorylating aspartate and producing aspartyl phosphate.

2. Aspartate semialdehyde dehydrogenase catalyzes the NADPH-dependent reduction of aspartyl phosphate to yield aspartate semialdehyde.

3. 4-hydroxy-tetrahydrodipicolinate synthase adds a pyruvate group to the β-aspartyl-4-semialdehyde, and a water molecule is removed. This causes cyclization and gives rise to (2S,4S)-4-hydroxy-2,3,4,5-tetrahydrodipicolinate.

4. 4-hydroxy-tetrahydrodipicolinate reductase catalyzes the reduction of (2S,4S)-4-hydroxy-2,3,4,5-tetrahydrodipicolinate by NADPH to yield Δ'-piperideine-2,6-dicarboxylate (2,3,4,5-tetrahydrodipicolinate) and H_2O.

5. Tetrahydrodipicolinate acyltransferase catalyzes the acetylation reaction that results in ring opening and yields N-acetyl α-amino-ε-ketopimelate.

6. N-succinyl-α-amino-ε-ketopimelate-glutamate aminotransaminase catalyzes the transamination reaction that removes the keto group of N-acetyl α-amino-ε-ketopimelate and replaces it with an amino group to yield N-succinyl-L-diaminopimelate.

7. N-acyldiaminopimelate deacylase catalyzes the deacylation of N-succinyl-L-diaminopimelate to yield L,L-diaminopimelate.

8. DAP epimerase catalyzes the conversion of L,L-diaminopimelate to the meso form of L,L-diaminopimelate.

9. DAP decarboxylase catalyzes the removal of the carboxyl group, yielding L-lysine.

The Serine Family of Amino Acids

The serine family of amino acid includes: serine, cysteine, and glycine. Most microorganisms and plants obtain the sulfur for synthesizing methionine from the amino acid cysteine. Furthermore, the conversion of serine to glycine provides the carbons needed for the biosynthesis of the methionine and histidine.

During serine biosynthesis, the enzyme phosphoglycerate dehydrogenase catalyzes the initial reaction that oxidizes 3-phospho-D-glycerate to yield 3-phosphonooxypyruvate. The following reaction is catalyzed by the enzyme phosphoserine aminotransferase, which transfers an amino group from glutamate onto 3-phosphonooxypyruvate to yield L-phosphoserine. The final step is catalyzed by the enzyme phosphoserine phosphatase, which dephosphorylates L-phosphoserine to yield L-serine.

There are two known pathways for the biosynthesis of glycine. Organisms that use ethanol and acetate as the major carbon source utilize the glyconeogenic pathway to synthesize glycine. The other pathway of glycine biosynthesis is known as the glycolytic pathway. This pathway converts serine synthesized from the intermediates of glycolysis to glycine. In the glycolytic pathway, the enzyme serine hydroxymethyltransferase catalyzes the cleavage of serine to yield glycine and transfers the cleaved carbon group of serine onto tetrahydrofolate, forming 5,10-methylene-tetrahydrofolate.

Cysteine biosynthesis is a two-step reaction that involves the incorporation of inorganic sulfur. In microorganisms and plants, the enzyme serine acetyltransferase catalyzes the transfer of acetyl group from acetyl-CoA onto L-serine to yield O-acetyl-L-serine. The following reaction step, catalyzed by the enzyme O-acetyl serine (thiol) lyase, replaces the acetyl group of O-acetyl-L-serine with sulfide to yield cysteine.

The Aspartate Family of Amino Acids

The aspartate family of amino acids includes: threonine, lysine, methionine, isoleucine, and aspartate. Lysine and isoleucine are considered part of the aspartate family even though part of their carbon skeleton is derived from pyruvate. In the case of methionine, the methyl carbon is derived from serine and the sulfur group, but in most organisms, it is derived from cysteine.

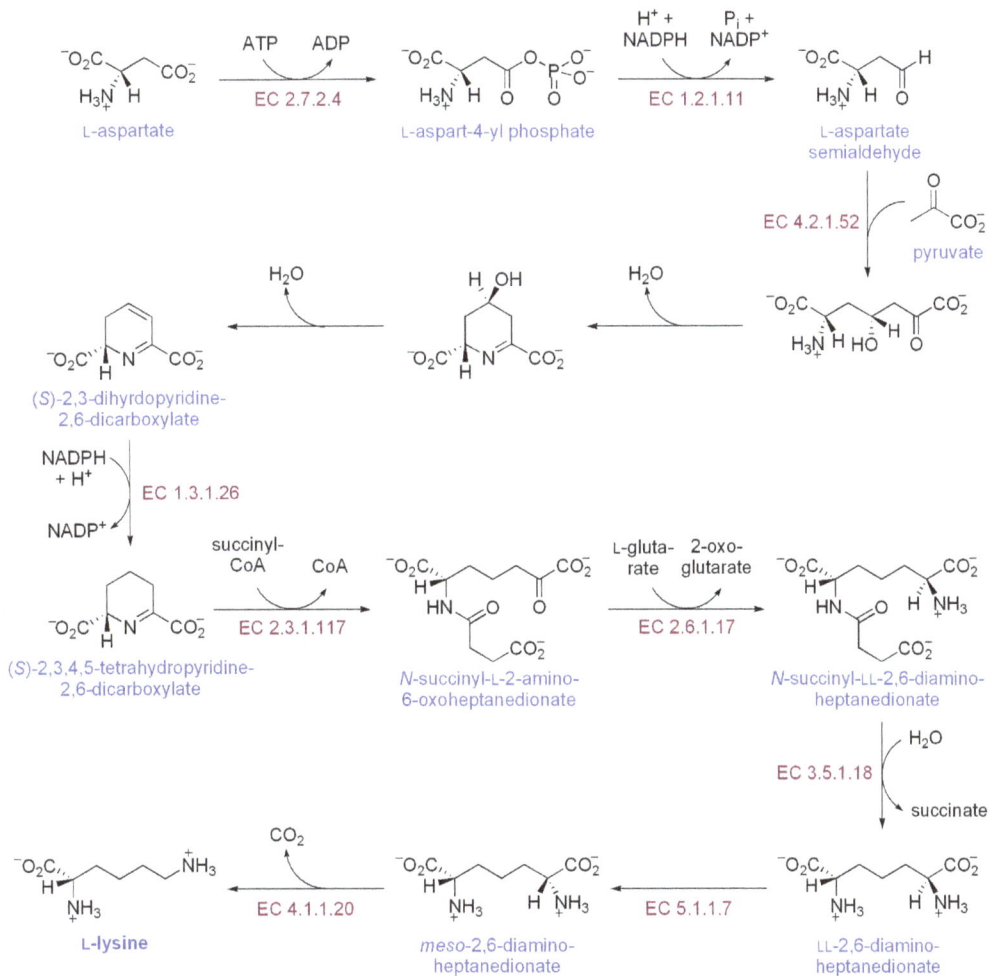

The diaminopimelic acid lysine biosynthetic pathway

The biosynthesis of aspartate is a one step reaction that is catalyzed by a single enzyme. The enzyme aspartate aminotransferase catalyzes the transfer of an amino group from aspartate onto α-ketoglutarate to yield glutamate and oxaloacetate. Asparagine is synthesized by an ATP-dependent addition of an amino group onto aspartate; asparagine synthetase catalyzes the addition of nitrogen from glutamine or soluble ammonia to aspartate to yield asparagine.

The diaminopimelic acid biosynthetic pathway of lysine belongs to the aspartate family of amino acids. This pathway involves nine enzyme-catalyzed reactions that convert aspartate to lysine.

1. Aspartate kinase catalyzes the initial step in the diaminopimelic acid pathway by transferring a phosphoryl from ATP onto the carboxylate group of aspartate, which yields aspartyl-β-phosphate.

2. Aspartate-semialdehyde dehydrogenase catalyzes the reduction reaction by dephosphorylation of aspartyl-β-phosphate to yield aspartate-β-semialdehyde.

3. Dihydrodipicolinate synthase catalyzes the condensation reaction of aspartate-β-semialdehyde with pyruvate to yield dihydrodipicolinic acid.

4. 4-hydroxy-tetrahydrodipicolinate reductase catalyzes the reduction of dihydrodipicolinic acid to yield tetrahydrodipicolinic acid.

5. Tetrahydrodipicolinate N-succinyltransferase catalyzes the transfer of a succinyl group from succinyl-CoA on to tetrahydrodipicolinic acid to yield N-succinyl-L-2,6-diaminoheptanedioate.

6. N-succinyldiaminopimelate aminotransferase catalyzes the transfer of an amino group from glutamate onto N-succinyl-L-2,6-diaminoheptanedioate to yield N-succinyl-L,L-diaminopimelic acid.

7. Succinyl-diaminopimelate desuccinylase catalyzes the removal of acyl group from N-succinyl-L,L-diaminopimelic acid to yield L,L-diaminopimelic acid.

8. Diaminopimelate epimerase catalyzes the inversion of the α-carbon of L,L-diaminopimelic acid to yield meso-diaminopimelic acid.

9. Siaminopimelate decarboxylase catalyzes the final step in lysine biosynthesis that removes the carbon dioxide group from meso-diaminopimelic acid to yield L-lysine.

Proteins

Protein synthesis occurs via a process called translation. During translation, genetic material called mRNA is read by ribosomes to generate a protein polypeptide chain. This process requires transfer RNA (tRNA) which serves as an adaptor by binding amino acids on one end and interacting with mRNA at the other end; the latter pairing between the tRNA and mRNA ensures that the correct amino acid is added to the chain. Protein synthesis occurs in three phases: initiation, elongation, and termination. Prokaryotic translation differs from eukaryotic translation; however, this section will mostly focus on the commonalities between the two organisms.

Peptide Synthesis

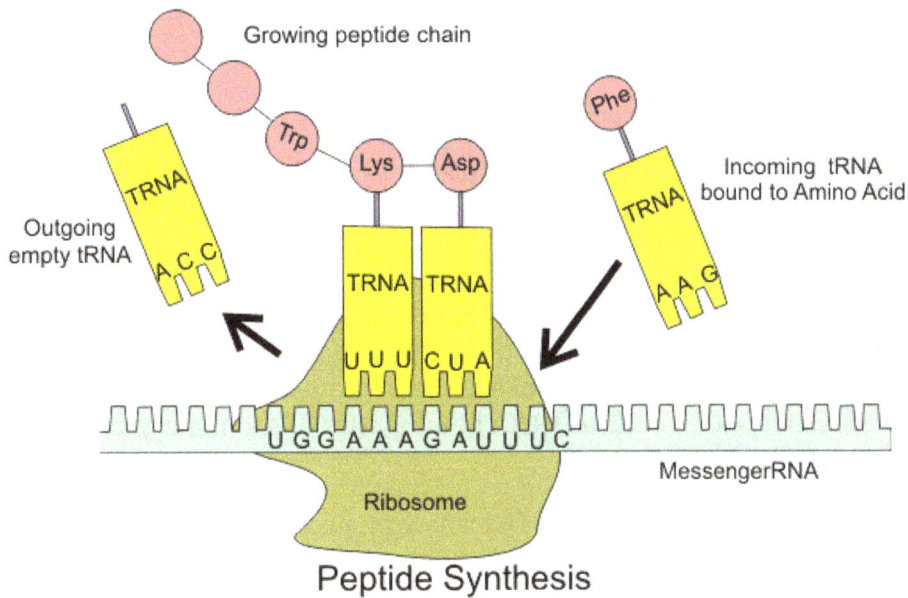

The tRNA anticodon interacts with the mRNA codon in order to bind an amino acid to growing polypeptide chain.

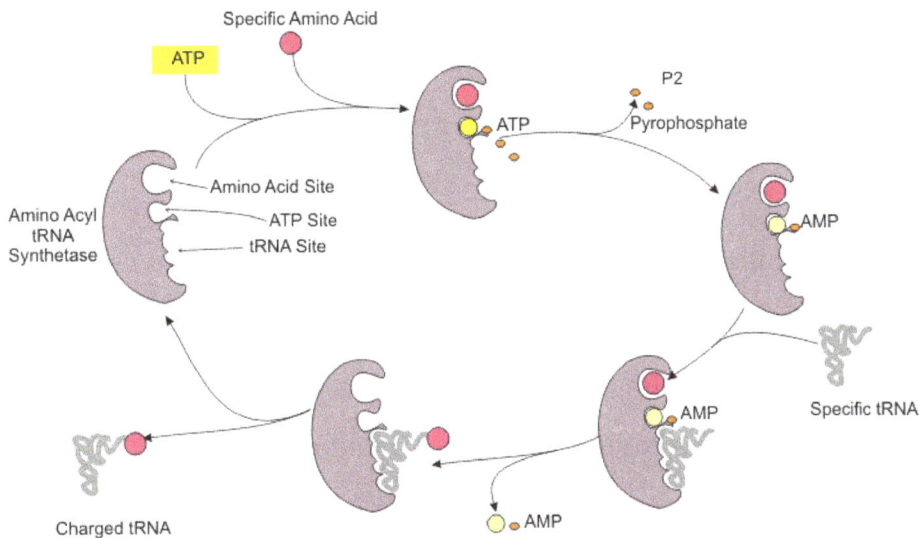

Frank Boumphrey M.D.
2009

Loading a tRNA molecule
The process of tRNA charging

Additional Background

Before translation can begin, the process of binding a specific amino acid to its corresponding tRNA must occur. This reaction, called tRNA charging, is catalyzed by aminoacyl tRNA synthetase. A specific tRNA synthetase is responsible for recognizing and charging a particular amino acid. Furthermore, this enzyme has special discriminator regions to ensure the correct binding between tRNA and its cognate amino acid. The first step for joining an amino acid to its corresponding tRNA is the formation of aminoacyl-AMP:

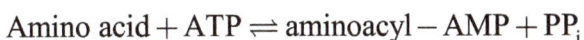

$$\text{Amino acid} + \text{ATP} \rightleftharpoons \text{aminoacyl} - \text{AMP} + \text{PP}_i$$

This is followed by the transfer of the aminoacyl group from aminoacyl-AMP to a tRNA molecule. The resulting molecule is aminoacyl-tRNA:

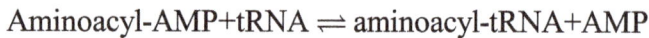

$$\text{Aminoacyl-AMP} + \text{tRNA} \rightleftharpoons \text{aminoacyl-tRNA} + \text{AMP}$$

The combination of these two steps, both of which are catalyzed by aminoacyl tRNA synthetase, produces a charged tRNA that is ready to add amino acids to the growing polypeptide chain.

In addition to binding an amino acid, tRNA has a three nucleotide unit called an anticodon that base pairs with specific nucleotide triplets on the mRNA called codons; codons encode a specific amino acid. This interaction is possible thanks to the ribosome, which serves as the site for protein synthesis. The ribosome possesses three tRNA binding sites: the aminoacyl site (A site), the peptidyl site (P site), and the exit site (E site).

There are numerous codons within an mRNA transcript, and it is very common for an amino acid to be specified by more than one codon; this phenomenon is called degeneracy. In all, there are 64 codons, 61 of each code for one of the 20 amino acids, while the remaining codons specify chain termination.

Translation in Steps

As previously mentioned, translation occurs in three phases: initiation, elongation, and termination.

Step 1: Initiation

The completion of the initiation phase is dependent on the following three events:

1. The recruitment of the ribosome to mRNA

2. The binding of a charged initiator tRNA into the P site of the ribosome

3. The proper alignment of the ribosome with mRNA's start codon

Step 2: Elongation

Following initiation, the polypeptide chain is extended via anticodon:codon interactions, with the ribosome adding amino acids to the polypeptide chain one at a time. The following steps must occur to ensure the correct addition of amino acids:

1. The binding of the correct tRNA into the A site of the ribosome

2. The formation of a peptide bond between the tRNA in the A site and the polypeptide chain attached to the tRNA in the P site

3. Translocation or advancement of the tRNA-mRNA complex by three nucleotides

Translocation "kicks off" the tRNA at the E site and shifts the tRNA from the A site into the P site, leaving the A site free for an incoming tRNA to add another amino acid.

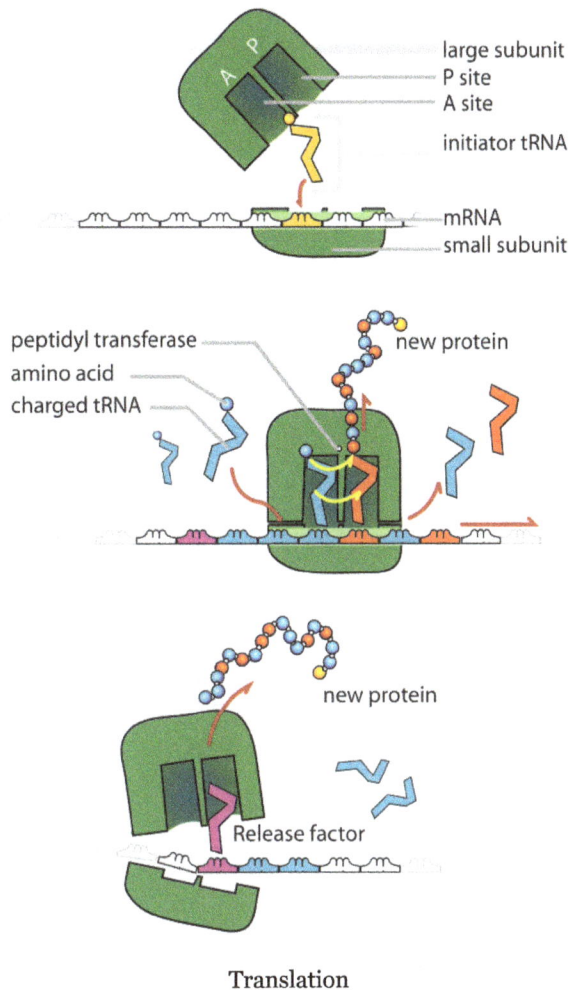

large subunit
P site
A site
initiator tRNA
mRNA
small subunit

peptidyl transferase
amino acid
charged tRNA
new protein

new protein

Release factor

Translation

Step 3: Termination

The last stage of translation occurs when a stop codon enters the A site. Then, the following steps occur:

1. The recognition of codons by release factors, which causes the hydrolysis of the polypeptide chain from the tRNA located in the P site

2. The release of the polypeptide chain

3. The dissociation and "recycling" of the ribosome for future translation processes

A summary table of the key players in translation is found below:

Key players in Translation	Translation Stage	Purpose
tRNA synthetase	before initiation	Responsible for tRNA charging
mRNA	initiation, elongation, termination	Template for protein synthesis; contains regions named codons which encode amino acids

| tRNA | initiation, elongation, termination | Binds ribosomes sites A, P, E; anticodon base pairs with mRNA codon to ensure that the correct amino acid is incorporated into the growing polypeptide chain |
| ribosome | initiation, elongation, termination | Directs protein synthesis and catalyzes the formation of the peptide bond |

Diseases Associated with Macromolecule Deficiency

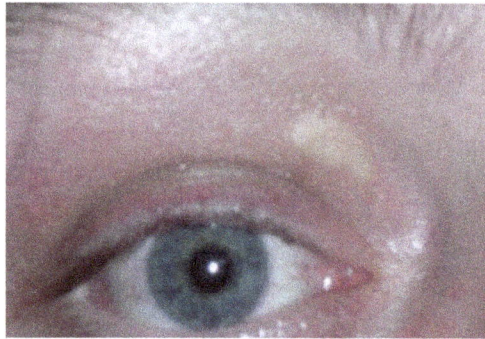

Familial hypercholesterolemia causes cholesterol deposits

Errors in biosynthetic pathways can have deleterious consequences including the malformation of macromolecules or the underproduction of functional molecules. Below are examples that illustrate the disruptions that occur due to these inefficiencies.

- Familial hypercholesterolemia: this disorder is characterized by the absence of functional receptors for LDL. Deficiencies in the formation of LDL receptors may cause faulty receptors which disrupt the endocytic pathway, inhibiting the entry of LDL into the liver and other cells. This causes a buildup of LDL in the blood plasma, which results in atherosclerotic plaques that narrow arteries and increase the risk of heart attacks.

- Lesch-Nyhan syndrome: this genetic disease is characterized by self- mutilation, mental deficiency, and gout. It is caused by the absence of hypoxanthine-guanine phosphoribosyltransferase, which is a necessary enzyme for purine nucleotide formation. The lack of enzyme reduces the level of necessary nucleotides and causes the accumulation of biosynthesis intermediates, which results in the aforementioned unusual behavior.

- Severe combined immunodeficiency (SCID): SCID is characterized by a loss of T cells. Shortage of these immune system components increases the susceptibility to infectious agents because the affected individuals cannot develop immunological memory. This immunological disorder results from a deficiency in adenosine deanimase activity, which causes a buildup of dATP. These dATP molecules then inhibit ribonucleotide reductase, which prevents of DNA synthesis.

- Huntington's disease: this neurological disease is caused from errors that occur during DNA synthesis. These errors or mutations lead to the expression of a mutant huntingtin protein, which contains repetitive glutamine residues that are encoded by expanding CAG trinucleotide repeats in the gene. Huntington's disease is characterized by neuronal loss and gliosis. Symptoms of the disease include: movement disorder, cognitive decline, and behavioral disorder.

Catalysis

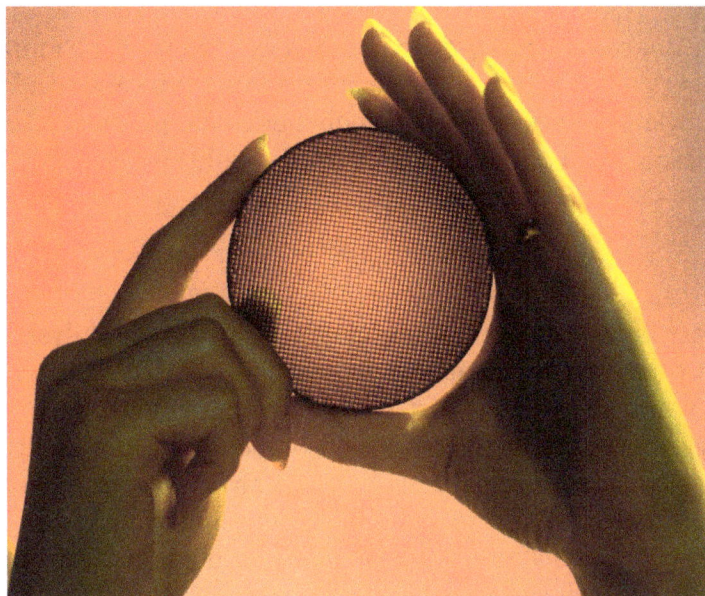

An air filter that utilizes low-temperature oxidation catalyst used to convert carbon monoxide to less toxic carbon dioxide at room temperature. It can also remove formaldehyde from the air.

Catalysis is the increase in the rate of a chemical reaction due to the participation of an additional substance called a catalyst. With a catalyst, reactions occur faster and require less activation energy. Because catalysts are not consumed in the catalyzed reaction, they can continue to catalyze the reaction of further quantities of reactant. Often only tiny amounts are required.

Technical Perspective

In the presence of a catalyst, less free energy is required to reach the transition state, but the total free energy from reactants to products does not change. A catalyst may participate in multiple chemical transformations. The effect of a catalyst may vary due to the presence of other substances known as inhibitors or poisons (which reduce the catalytic activity) or promoters (which increase the activity). The opposite of a catalyst, a substance that reduces the rate of a reaction, is an inhibitor.

Catalyzed reactions have a lower activation energy (rate-limiting free energy of activation) than the corresponding uncatalyzed reaction, resulting in a higher reaction rate at the same temperature and for the same reactant concentrations. However, the detailed mechanics of catalysis is complex. Catalysts may affect the reaction environment favorably, or bind to the reagents to polarize bonds, e.g. acid catalysts for reactions of carbonyl compounds, or form specific intermediates that are not produced naturally, such as osmate esters in osmium tetroxide-catalyzed dihydroxylation of alkenes, or cause dissociation of reagents to reactive forms, such as chemisorbed hydrogen in catalytic hydrogenation.

Kinetically, catalytic reactions are typical chemical reactions; i.e. the reaction rate depends on the frequency of contact of the reactants in the rate-determining step. Usually, the catalyst participates in this slowest step, and rates are limited by amount of catalyst and its "activity". In heterogeneous catalysis, the diffusion of reagents to the surface and diffusion of products from the surface can be rate deter-mining. A nanomaterial-based catalyst is an example of a heterogeneous catalyst. Analogous events associated with substrate binding and product dissociation apply to homogeneous catalysts.

Although catalysts are not consumed by the reaction itself, they may be inhibited, deactivated, or destroyed by secondary processes. In heterogeneous catalysis, typical secondary processes include coking where the catalyst becomes covered by polymeric side products. Additionally, heteroge-neous catalysts can dissolve into the solution in a solid–liquid system or sublimate in a solid–gas system.

Background

The production of most industrially important chemicals involves catalysis. Similarly, most bio-chemically significant processes are catalysed. Research into catalysis is a major field in applied science and involves many areas of chemistry, notably organometallic chemistry and materials science. Catalysis is relevant to many aspects of environmental science, e.g. the catalytic converter in automobiles and the dynamics of the ozone hole. Catalytic reactions are preferred in environ-mentally friendly green chemistry due to the reduced amount of waste generated, as opposed to stoichiometric reactions in which all reactants are consumed and more side products are formed. Many transition metals and transition metal complexes are used in catalysis as well. Catalysts called enzymes are important in biology.

A catalyst works by providing an alternative reaction pathway to the reaction product. The rate of the reaction is increased as this alternative route has a lower activation energy than the reaction route not mediated by the catalyst. The disproportionation of hydrogen peroxide creates water and oxygen, as shown below.

$$2\,H_2O_2 \rightarrow 2\,H_2O + O_2$$

This reaction is preferable in the sense that the reaction products are more stable than the starting material, though the uncatalysed reaction is slow. In fact, the decomposition of hydrogen peroxide is so slow that hydrogen peroxide solutions are commercially available. This reaction is strongly affected by catalysts such as manganese dioxide, or the enzyme peroxidase in organisms. Upon the addition of a small amount of manganese dioxide, the hydrogen peroxide reacts rapidly. This effect is readily seen by the effervescence of oxygen. The manganese dioxide is not consumed in the reaction, and thus may be recovered unchanged, and re-used indefinitely. Accordingly, manganese dioxide *catalyses* this reaction.

General Principles

Units

Catalytic activity is usually denoted by the symbol z and measured in mol/s, a unit which was called katal and defined the SI unit for catalytic activity since 1999. Catalytic activity is not a kind of reaction rate, but a property of the catalyst under certain conditions, in relation to a specific chemical reaction. Catalytic activity of one katal (Symbol 1 kat = 1 mol/s) of a catalyst means an amount of that catalyst (substance, in Mol) that leads to a net reaction of one Mol per second of the reactants to the resulting reagents or other outcome which was intended for this chemical reaction. A catalyst may and usually will have different catalytic activity for distinct reactions. There are further derived SI units related to catalytic activity.

Typical Mechanism

Catalysts generally react with one or more reactants to form intermediates that subsequently give the final reaction product, in the process regenerating the catalyst. The following is a typical reaction scheme, where C represents the catalyst, X and Y are reactants, and Z is the product of the reaction of X and Y:

$$X + C \rightarrow XC \tag{1}$$

$$Y + XC \rightarrow XYC \tag{2}$$

$$XYC \rightarrow CZ \tag{3}$$

$$CZ \rightarrow C + Z \tag{4}$$

Although the catalyst is consumed by reaction 1, it is subsequently produced by reaction 4, so it does not occur in the overall reaction equation:

$$X + Y \rightarrow Z$$

As a catalyst is regenerated in a reaction, often only small amounts are needed to increase the rate of the reaction. In practice, however, catalysts are sometimes consumed in secondary processes.

The catalyst does usually appear in the rate equation. For example, if the rate-determining step in the above reaction scheme is the first step
$X + C \rightarrow XC$, the catalyzed reaction will be second order with rate equation $v = k_{cat}[X][C]$, which is proportional to the catalyst concentration [C]. However [C] remains constant during the reaction so that the catalyzed reaction is pseudo-first order: $v = k_{obs}[X]$, where $k_{obs} = k_{cat}[C]$.

As an example of a detailed mechanism at the microscopic level, in 2008 Danish researchers first revealed the sequence of events when oxygen and hydrogen combine on the surface of titanium dioxide (TiO_2, or *titania*) to produce water. With a time-lapse series of scanning tunneling microscopy images, they determined the molecules undergo adsorption, dissociation and diffusion before reacting. The intermediate reaction states were: HO_2, H_2O_2, then H_3O_2 and the final reaction product (water molecule dimers), after which the water molecule desorbs from the catalyst surface.

Reaction Energetics

Catalysts work by providing an (alternative) mechanism involving a different transition state and lower activation energy. Consequently, more molecular collisions have the energy needed to reach the transition state. Hence, catalysts can enable reactions that would otherwise be blocked or slowed by a kinetic barrier. The catalyst may increase reaction rate or selectivity, or enable the reaction at lower temperatures. This effect can be illustrated with an energy profile diagram.

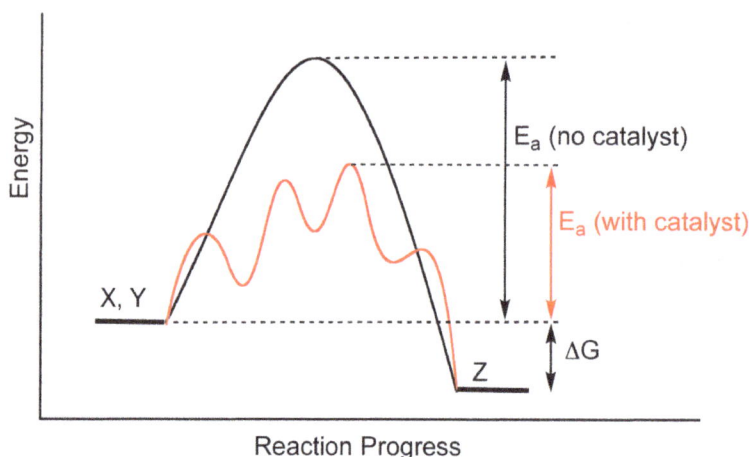

Generic potential energy diagram showing the effect of a catalyst in a hypothetical exothermic chemical reaction X + Y to give Z. The presence of the catalyst opens a different reaction pathway (shown in red) with a lower activation energy. The final result and the overall thermodynamics are the same.

In the catalyzed elementary reaction, catalysts do not change the extent of a reaction: they have no effect on the chemical equilibrium of a reaction because the rate of both the forward and the reverse reaction are both affected. The second law of thermodynamics describes why a catalyst does not change the chemical equilibrium of a reaction. Suppose there was such a catalyst that shifted an equilibrium. Introducing the catalyst to the system would result in a reaction to move to the new equilibrium, producing energy. Production of energy is a neces-sary result since reactions are spontaneous only if Gibbs free energy is produced, and if there is no energy barrier, there is no need for a catalyst. Then, removing the catalyst would also result in reaction, producing energy; i.e. the addition and its reverse process, removal, would both produce energy. Thus, a catalyst that could change the equilibrium would be a perpetual motion machine, a contradiction to the laws of thermodynamics.

If a catalyst does change the equilibrium, then it must be consumed as the reaction proceeds, and thus it is also a reactant. Illustrative is the base-catalysed hydrolysis of esters, where the produced carboxylic acid immediately reacts with the base catalyst and thus the reaction equilibrium is shifted towards hydrolysis.

The SI derived unit for measuring the catalytic activity of a catalyst is the katal, which is moles per second. The productivity of a catalyst can be described by the turn over number (or TON) and the catalytic activity by the *turn over frequency* (TOF), which is the TON per time unit. The biochemical equivalent is the enzyme unit. For more information on the efficiency of enzymatic catalysis.

The catalyst stabilizes the transition state more than it stabilizes the starting material. It decreases the kinetic barrier by decreasing the *difference* in energy between starting material and transition state. It does not change the energy difference between starting materials and products (thermodynamic barrier), or the available energy (this is provided by the environment as heat or light).

Materials

The chemical nature of catalysts is as diverse as catalysis itself, although some generalizations can be made. Proton acids are probably the most widely used catalysts, especially for the many reactions involving water, including hydrolysis and its reverse. Multifunctional solids often are catalytically active, e.g. zeolites, alumina, higher-order oxides, graphitic carbon, nanoparticles, nanodots, and facets of bulk materials. Transition metals are often used to catalyze redox reactions (oxidation, hydrogenation). Examples are nickel, such as Raney nickel for hydrogenation, and vanadium(V) oxide for oxidation of sulfur dioxide into sulfur trioxide by the so-called contact process. Many catalytic processes, especially those used in organic synthesis, require "late transition metals", such as palladium, platinum, gold, ruthenium, rhodium, or iridium.

Some so-called catalysts are really precatalysts. Precatalysts convert to catalysts in the reaction. For example, Wilkinson's catalyst $RhCl(PPh_3)_3$ loses one triphenylphosphine ligand before entering the true catalytic cycle. Precatalysts are easier to store but are easily activated in situ. Because of this preactivation step, many catalytic reactions involve an induction period.

Chemical species that improve catalytic activity are called co-catalysts (cocatalysts) or promotors in cooperative catalysis.

Types

Catalysts can be heterogeneous or homogeneous, depending on whether a catalyst exists in the same phase as the substrate. Biocatalysts (enzymes) are often seen as a separate group.

Heterogeneous Catalysts

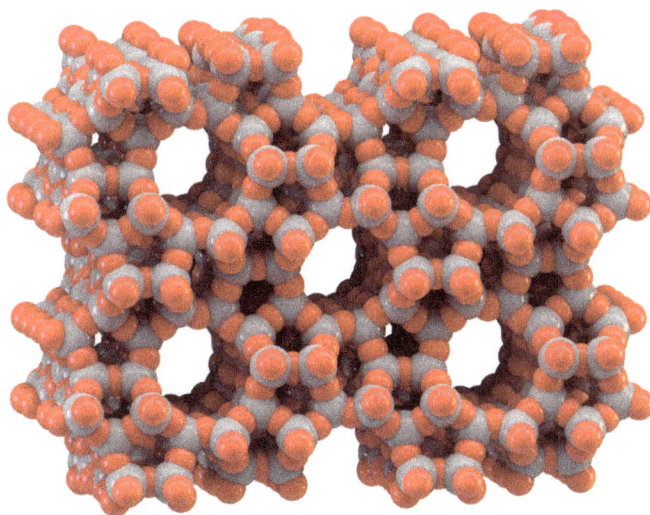

The microporous molecular structure of the zeolite ZSM-5 is exploited in catalysts used in refineries

Heterogeneous catalysts act in a different phase than the reactants. Most heterogeneous catalysts are solids that act on substrates in a liquid or gaseous reaction mixture. Diverse mechanisms for reactions on surfaces are known, depending on how the adsorption takes place (Langmuir-Hinshelwood, Eley-Rideal, and Mars-van Krevelen). The total surface area of solid has an important effect on the reaction rate. The smaller the catalyst particle size, the larger the surface area for a given mass of particles.

Zeolites are extruded as pellets for easy handling in catalytic reactors.

A heterogeneous catalyst has active sites, which are the atoms or crystal faces where the reaction actually occurs. Depending on the mechanism, the active site may be either a planar exposed metal surface, a crystal edge with imperfect metal valence or a complicated combination of the two. Thus, not only most of the volume, but also most of the surface of a heterogeneous catalyst may be catalytically inactive. Finding out the nature of the active site requires technically challenging research. Thus, empirical research for finding out new metal combinations for catalysis continues.

For example, in the Haber process, finely divided iron serves as a catalyst for the synthesis of ammonia from nitrogen and hydrogen. The reacting gases adsorb onto active sites on the iron particles. Once physically adsorbed, the reagents undergo chemisorption that results in dissociation into adsorbed atomic species, and new bonds between the resulting fragments form in part due to their close proximity. In this way the particularly strong triple bond in nitrogen is broken, which would be extremely uncommon in the gas phase due to its high activation energy. Thus, the activation energy of the overall reaction is lowered, and the rate of reaction increases. Another place where a heterogeneous catalyst is applied is in the oxidation of sulfur dioxide on vanadium(V) oxide for the production of sulfuric acid.

Heterogeneous catalysts are typically "supported," which means that the catalyst is dispersed on a second material that enhances the effectiveness or minimizes their cost. Supports prevent or reduce agglomeration and sintering of the small catalyst particles, exposing more surface area, thus catalysts have a higher specific activity (per gram) on a support. Sometimes the support is merely a surface on which the catalyst is spread to increase the surface area. More often, the support and the catalyst interact, affecting the catalytic reaction. Supports are porous materials with a high surface area, most commonly alumina, zeolites or various kinds of activated carbon. Specialized supports include silicon dioxide, titanium dioxide, calcium carbonate, and barium sulfate.

Electrocatalysts

In the context of electrochemistry, specifically in fuel cell engineering, various metal-containing catalysts are used to enhance the rates of the half reactions that comprise the fuel cell. One common type of fuel cell electrocatalyst is based upon nanoparticles of platinum that are supported on slightly larger carbon particles. When in contact with one of the electrodes in a fuel cell, this platinum increases the rate of oxygen reduction either to water, or to hydroxide or hydrogen peroxide.

Homogeneous Catalysts

Homogeneous catalysts function in the same phase as the reactants, but the mechanistic principles invoked in heterogeneous catalysis are generally applicable. Typically homogeneous catalysts are dissolved in a solvent with the substrates. One example of homogeneous catalysis involves the influence of H^+ on the esterification of carboxylic acids, such as the formation of methyl acetate from acetic acid and methanol. For inorganic chemists, homogeneous catalysis is often synonymous with organometallic catalysts.

Organocatalysis

Whereas transition metals sometimes attract most of the attention in the study of catalysis, small organic molecules without metals can also exhibit catalytic properties, as is apparent from the fact that many enzymes lack transition metals. Typically, organic catalysts require a higher loading (amount of catalyst per unit amount of reactant, expressed in mol% amount of substance) than transition metal(-ion)-based catalysts, but these catalysts are usually commercially available in bulk, helping to reduce costs. In the early 2000s, these organocatalysts were considered "new generation" and are competitive to traditional metal(-ion)-containing catalysts. Organocatalysts are supposed to operate akin to metal-free enzymes utilizing, e.g., non-covalent interactions such as hydrogen bonding. The discipline organocatalysis is divided in the application of covalent (e.g., proline, DMAP) and non-covalent (e.g., thiourea organocatalysis) organocatalysts referring to the preferred catalyst-substrate binding and interaction, respectively.

Enzymes and Biocatalysts

In biology, enzymes are protein-based catalysts in metabolism and catabolism. Most biocatalysts are enzymes, but other non-protein-based classes of biomolecules also exhibit catalytic properties including ribozymes, and synthetic deoxyribozymes.

Biocatalysts can be thought of as intermediate between homogeneous and heterogeneous catalysts, although strictly speaking soluble enzymes are homogeneous catalysts and membrane-bound enzymes are heterogeneous. Several factors affect the activity of enzymes (and other catalysts) including temperature, pH, concentration of enzyme, substrate, and products. A particularly important reagent in enzymatic reactions is water, which is the product of many bond-forming reactions and a reactant in many bond-breaking processes.

In biocatalysis, enzymes are employed to prepare many commodity chemicals including high-fructose corn syrup and acrylamide.

Some monoclonal antibodies whose binding target is a stable molecule which resembles the transition state of a chemical reaction can function as weak catalysts for that chemical reaction by lowering its activation energy. Such catalytic antibodies are sometimes called "abzymes".

Nanocatalysts

Nanocatalysts are nanomaterials with catalytic activities. They have been extensively explored for wide range of applications. Among them, the nanocatalysts with enzyme mimicking activities are collectively called as nanozymes.

Tandem Catalysis

In tandem catalysis two or more different catalysts are coupled in a one-pot reaction.

Autocatalysis

In autocatalysis the catalyst *is* a product of the overall reaction, in contrast to all other types of catalysis considered in this article. The simplest example of autocatalysis is a reaction of type A + B → 2 B, in one or in several steps. The overall reaction is just A → B, so that B is a product. But since B is also a reactant, it may be present in the rate equation and affect the reaction rate. As the reaction proceeds, the concentration of B increases and can accelerate the reaction as a catalyst. In effect, the reaction accelerates itself or is autocatalyzed.

A real example is the hydrolysis of an ester such as aspirin to a carboxylic acid and an alcohol. In the absence of added acid catalysts, the carboxylic acid product catalyzes the hydrolysis.

Significance

Left: Partially caramelised cube sugar, Right: burning cube sugar with ash as catalyst

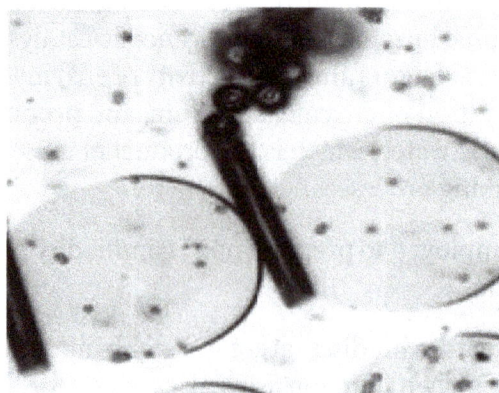

A Ti-Cr-Pt tube (~40 μm long) releases oxygen bubbles when immersed in hydrogen peroxide (via catalytic decomposition), forming a micropump.

Estimates are that 90% of all commercially produced chemical products involve catalysts at some stage in the process of their manufacture. In 2005, catalytic processes generated about $900 billion in products worldwide. Catalysis is so pervasive that subareas are not readily classified. Some areas of particular concentration are surveyed below.

Energy Processing

Petroleum refining makes intensive use of catalysis for alkylation, catalytic cracking (breaking long-chain hydrocarbons into smaller pieces), naphtha reforming and steam reforming (conversion of hydrocarbons into synthesis gas). Even the exhaust from the burning of fossil fuels is treated via catalysis: Catalytic converters, typically composed of platinum and rhodium, break down some of the more harmful byproducts of automobile exhaust.

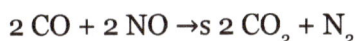

$$2\, CO + 2\, NO \rightarrow s\, 2\, CO_2 + N_2$$

With regard to synthetic fuels, an old but still important process is the Fischer-Tropsch synthesis of hydrocarbons from synthesis gas, which itself is processed via water-gas shift reactions, catalysed by iron. Biodiesel and related biofuels require processing via both inorganic and biocatalysts.

Fuel cells rely on catalysts for both the anodic and cathodic reactions.

Catalytic heaters generate flameless heat from a supply of combustible fuel.

Bulk Chemicals

Some of the largest-scale chemicals are produced via catalytic oxidation, often using oxygen. Examples include nitric acid (from ammonia), sulfuric acid (from sulfur dioxide to sulfur trioxide by the chamber process), terephthalic acid from p-xylene, and acrylonitrile from propane and ammonia.

Many other chemical products are generated by large-scale reduction, often via hydrogenation. The largest-scale example is ammonia, which is prepared via the Haber process from nitrogen. Methanol is prepared from carbon monoxide.

Bulk polymers derived from ethylene and propylene are often prepared via Ziegler-Natta catalysis. Polyesters, polyamides, and isocyanates are derived via acid-base catalysis.

Most carbonylation processes require metal catalysts, examples include the Monsanto acetic acid process and hydroformylation.

Fine Chemicals

Many fine chemicals are prepared via catalysis; methods include those of heavy industry as well as more specialized processes that would be prohibitively expensive on a large scale. Examples include the Heck reaction, and Friedel-Crafts reactions.

Because most bioactive compounds are chiral, many pharmaceuticals are produced by enantioselective catalysis (catalytic asymmetric synthesis).

Food Processing

One of the most obvious applications of catalysis is the hydrogenation (reaction with hydrogen gas) of fats using nickel catalyst to produce margarine. Many other foodstuffs are prepared via biocatalysis.

Environment

Catalysis impacts the environment by increasing the efficiency of industrial processes, but catalysis also plays a direct role in the environment. A notable example is the catalytic role of chlorine free radicals in the breakdown of ozone. These radicals are formed by the action of ultraviolet radiation on chlorofluorocarbons (CFCs).

$$Cl^{\cdot} + O_3 \rightarrow ClO^{\cdot} + O_2$$

$$ClO^{\cdot} + O^{\cdot} \rightarrow Cl^{\cdot} + O_2$$

History

In a general sense, anything that increases the rate of a process is a "catalyst", a term derived from Greek καταλύειν, meaning "to annul," or "to untie," or "to pick up." The concept of catalysis was invented by chemist Elizabeth Fulhame and described in a 1794 book, based on her novel work in oxidation-reduction experiments. The term *catalysis* was later used by Jöns Jakob Berzelius in 1835 to describe reactions that are accelerated by substances that remain unchanged after the reaction. Fulhame, who predated Berzelius, did work with water as opposed to metals in her reduction experiments. Other 18th century chemists who worked in catalysis were Eilhard Mitscherlich who referred to it as *contact* processes, and Johann Wolfgang Döbereiner who spoke of *contact action*. He developed Döbereiner's lamp, a lighter based on hydrogen and a platinum sponge, which became a commercial success in the 1820s that lives on today. Humphry Davy discovered the use of platinum in catalysis. In the 1880s, Wilhelm Ostwald at Leipzig University started a systematic investigation into reactions that were catalyzed by the presence of acids and bases, and found that chemical reactions occur at finite rates and that these rates can be used to determine the strengths of acids and bases. For this work, Ostwald was awarded the 1909 Nobel Prize in Chemistry.

Inhibitors, Poisons and Promoters

Substances that reduce the action of catalysts are called catalyst inhibitors if reversible, and catalyst poisons if irreversible. Promoters are substances that increase the catalytic activity, even though they are not catalysts by themselves.

Inhibitors are sometimes referred to as "negative catalysts" since they decrease the reaction rate. However the term inhibitor is preferred since they do not work by introducing a reaction path with higher activation energy; this would not reduce the rate since the reaction would continue to occur by the non-catalyzed path. Instead they act either by deactivating catalysts, or by removing reaction intermediates such as free radicals.

The inhibitor may modify selectivity in addition to rate. For instance, in the reduction of alkynes to alkenes, a palladium (Pd) catalyst partly "poisoned" with lead(II) acetate ($Pb(CH_3CO_2)_2$) can be used. Without the deactivation of the catalyst, the alkene produced would be further reduced to alkane.

The inhibitor can produce this effect by, e.g., selectively poisoning only certain types of active sites. Another mechanism is the modification of surface geometry. For instance, in hydrogenation operations, large planes of metal surface function as sites of hydrogenolysis catalysis while sites catalyzing hydrogenation of unsaturates are smaller. Thus, a poison that covers surface randomly will tend to reduce the number of uncontaminated large planes but leave proportionally more smaller sites free, thus changing the hydrogenation vs. hydrogenolysis selectivity. Many other mechanisms are also possible.

Promoters can cover up surface to prevent production of a mat of coke, or even actively remove such material (e.g., rhenium on platinum in platforming). They can aid the dispersion of the catalytic material or bind to reagents.

Current Market

The global demand on catalysts in 2010 was estimated at approximately 29.5 billion USD. With the rapid recovery in automotive and chemical industry overall, the global catalyst market is expected to experience fast growth in the next years.

References

- Buffum, Burt C. (2008). Arid Agriculture; A Hand-Book for the Western Farmer and Stockman. Read Books. p. 232. ISBN 978-1-4086-6710-1.

- Grandin, Temple; Johnson, Catherine (2005). Animals in Translation. New York, New York: Scribner. pp. 69–71. ISBN 0-7432-4769-8.

- Jain, H. K.; Kharkwal, M. C. (2004). Plant breeding - Mendelian to molecular approaches. Boston, London, Dordecht: Kluwer Academic Publishers. ISBN 1-4020-1981-5.

- Kirpichnikov, V. S.; Ilyasov, I.; Shart, L. A.; Vikhman, A. A.; Ganchenko, M. V.; Ostashevsky, A. L.; Simonov, V. M.; Tikhonov, G. F.; Tjurin, V. V. (1993). "Selection of Krasnodar common carp (Cyprinus carpio L.) for resistance to dropsy: Principal results and prospects". Genetics in Aquaculture. p. 7. doi:10.1016/b978-0-444-81527-9.50006-3. ISBN 9780444815279.

- Pratt, Donald Voet, Judith G. Voet, Charlotte W. Fundamentals of biochemistry : life at the molecular level (4th ed.). Hoboken, NJ: Wiley. ISBN 978-0470547847.

- Cox, David L. Nelson, Michael M. (2008). Lehninger principles of biochemistry (5th ed.). New York: W.H. Freeman. ISBN 9780716771081.

- Vance, Dennis E.; Vance, Jean E. (2008). Biochemistry of lipids, lipoproteins and membranes (5th ed.). Amsterdam: Elsevier. ISBN 978-0444532190.

- Stryer, Jeremy M. Berg; John L. Tymoczko; Lubert (2007). Biochemistry (6. ed., 3. print. ed.). New York: Freeman. ISBN 978-0716787242.

- Siegel, George J. (1999). Basic neurochemistry : molecular, cellular and medical aspects (6. ed.). Philadelphia, Pa. [u.a.]: Lippincott Williams & Wilkins. ISBN 978-0397518203.

- Harris, J. Robin (2010). Cholesterol binding and cholesterol transport proteins : structure and function in health and disease. Dordrecht: Springer. ISBN 978-9048186211.

- Watson, James D.; et al. (2007). Molecular biology of the gene (6th ed.). San Francisco, Calif.: Benjamin Cummings. ISBN 978-0805395921

- .Broach, edited by Jeffrey N. Strathern, Elizabeth W. Jones, James R. (1981). The Molecular biology of the yeast Saccharomyces. Cold Spring Harbor, N.Y.: Cold Spring Harbor Laboratory. ISBN 0879691395.

- Geer, Gerald Karp ; responsible for the revision of chapter 15 Peter van der (2004). Cell and molecular biology : concepts and experiments (4th ed., Wiley International ed.). New York: J. Wiley & Sons. ISBN 978-0471656654.

4

Applications of Biotechnology

Biotechnology is best understood in confluence with the major topics listed in the following chapter. The applications of biotechnology covered in this chapter are cloning, genetic engineering, recombinant DNA and tissue engineering. The topics discussed in the chapter are of great importance to broaden the existing knowledge on biotechnology.

Cloning

Many organisms, including aspen trees, reproduce by cloning.

In biology, cloning is the process of producing similar populations of genetically identical individuals that occurs in nature when organisms such as bacteria, insects or plants reproduce asexually. Cloning in biotechnology refers to processes used to create copies of DNA fragments (molecular cloning), cells (cell cloning), or organisms. The term also refers to the production of multiple copies of a product such as digital media or software.

The term clone, invented by J. B. S. Haldane, is derived from the Ancient Greek word κλών *klōn*, "twig", referring to the process whereby a new plant can be created from a twig. In horticulture, the spelling *clon* was used until the twentieth century; the final *e* came into use to indicate the vowel is a "long o" instead of a "short o". Since the term entered the popular lexicon in a more general context, the spelling *clone* has been used exclusively.

In botany, the term lusus was traditionally used.

Natural Cloning

Cloning is a natural form of reproduction that has allowed life forms to spread for more than 50 thousand years. It is the reproduction method used by plants, fungi, and bacteria, and is also the way that clonal colonies reproduce themselves. Examples of these organisms include blueberry plants, hazel trees, the Pando trees, the Kentucky coffeetree, *Myricas*, and the American sweet-gum.

Molecular Cloning

Molecular cloning refers to the process of making multiple molecules. Cloning is commonly used to amplify DNA fragments containing whole genes, but it can also be used to amplify any DNA sequence such as promoters, non-coding sequences and randomly fragmented DNA. It is used in a wide array of biological experiments and practical applications ranging from genetic fingerprinting to large scale protein production. Occasionally, the term cloning is misleadingly used to refer to the identification of the chromosomal location of a gene associated with a particular phenotype of interest, such as in positional cloning. In practice, localization of the gene to a chromosome or genomic region does not necessarily enable one to isolate or amplify the relevant genomic sequence. To amplify any DNA sequence in a living organism, that sequence must be linked to an origin of replication, which is a sequence of DNA capable of directing the propagation of itself and any linked sequence. However, a number of other features are needed, and a variety of specialised cloning vectors (small piece of DNA into which a foreign DNA fragment can be inserted) exist that allow protein production, affinity tagging, single stranded RNA or DNA production and a host of other molecular biology tools.

Cloning of any DNA fragment essentially involves four steps

1. fragmentation - breaking apart a strand of DNA

2. ligation - gluing together pieces of DNA in a desired sequence

3. transfection - inserting the newly formed pieces of DNA into cells

4. screening/selection - selecting out the cells that were successfully transfected with the new DNA

Although these steps are invariable among cloning procedures a number of alternative routes can be selected; these are summarized as a *cloning strategy*.

Initially, the DNA of interest needs to be isolated to provide a DNA segment of suitable size. Subsequently, a ligation procedure is used where the amplified fragment is inserted into a vector (piece of DNA). The vector (which is frequently circular) is linearised using restriction enzymes, and incubated with the fragment of interest under appropriate conditions with an enzyme called DNA ligase. Following ligation the vector with the insert of interest is transfected into cells. A number of alternative techniques are available, such as chemical sensitivation of cells, electroporation, optical injection and biolistics. Finally, the transfected cells are cultured. As the aforementioned procedures are of particularly low efficiency, there is a need to identify the cells that have been successfully transfected with the vector construct containing the desired insertion sequence in the required orientation. Modern cloning vectors include selectable antibiotic resistance markers, which allow only cells in which the vector has been transfected, to grow. Additionally, the cloning vectors may contain colour selection markers, which provide blue/white screening (alpha-factor complementation) on X-gal medium. Nevertheless, these selection steps do not absolutely guarantee that the DNA insert is present in the cells obtained. Further investigation of the resulting colonies must be required to confirm that cloning was successful. This may be accomplished by means of PCR, restriction fragment analysis and/or DNA sequencing.

Cell Cloning

Cloning Unicellular Organisms

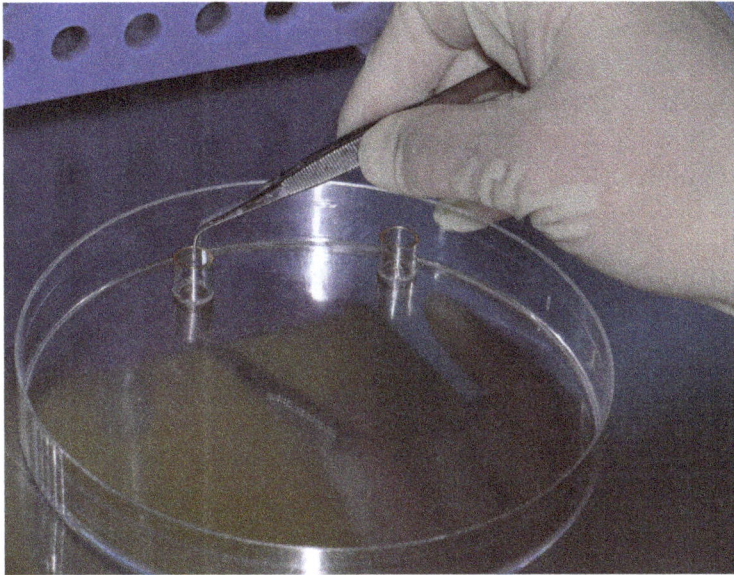

Cloning cell-line colonies using cloning rings

Cloning a cell means to derive a population of cells from a single cell. In the case of unicellular organisms such as bacteria and yeast, this process is remarkably simple and essentially only requires the inoculation of the appropriate medium. However, in the case of cell cultures from multi-cellular organisms, cell cloning is an arduous task as these cells will not readily grow in standard media.

A useful tissue culture technique used to clone distinct lineages of cell lines involves the use of cloning rings (cylinders). In this technique a single-cell suspension of cells that have been exposed to a mutagenic agent or drug used to drive selection is plated at high dilution to create isolated colonies, each arising from a single and potentially clonal distinct cell. At an early growth stage when colonies consist of only a few cells, sterile polystyrene rings (cloning rings), which have been dipped in grease, are placed over an individual colony and a small amount of trypsin is added. Cloned cells are collected from inside the ring and transferred to a new vessel for further growth.

Cloning Stem Cells

Somatic-cell nuclear transfer, known as SCNT, can also be used to create embryos for research or therapeutic purposes. The most likely purpose for this is to produce embryos for use in stem cell research. This process is also called "research cloning" or "therapeutic cloning." The goal is not to create cloned human beings (called "reproductive cloning"), but rather to harvest stem cells that can be used to study human development and to potentially treat disease. While a clonal human blastocyst has been created, stem cell lines are yet to be isolated from a clonal source.

Therapeutic cloning is achieved by creating embryonic stem cells in the hopes of treating diseases such as diabetes and Alzheimer's. The process begins by removing the nucleus (containing the DNA) from an egg cell and inserting a nucleus from the adult cell to be cloned. In the case of someone with Alzheimer's disease, the nucleus from a skin cell of that patient is placed into an empty egg. The reprogrammed cell begins to develop into an embryo because the egg reacts with the transferred nucleus. The embryo will become genetically identical to the patient. The embryo will then form a blastocyst which has the potential to form/become any cell in the body.

The reason why SCNT is used for cloning is because somatic cells can be easily acquired and cultured in the lab. This process can either add or delete specific genomes of farm animals. A key point to remember is that cloning is achieved when the oocyte maintains its normal functions and instead of using sperm and egg genomes to replicate, the oocyte is inserted into the donor's somatic cell nucleus. The oocyte will react on the somatic cell nucleus, the same way it would on sperm cells.

The process of cloning a particular farm animal using SCNT is relatively the same for all animals. The first step is to collect the somatic cells from the animal that will be cloned. The somatic cells could be used immediately or stored in the laboratory for later use. The hardest part of SCNT is removing maternal DNA from an oocyte at metaphase II. Once this has been done, the somatic nucleus can be inserted into an egg cytoplasm. This creates a one-cell embryo. The grouped somatic cell and egg cytoplasm are then introduced to an electrical current. This energy will hopefully allow the cloned embryo to begin development. The successfully developed embryos are then placed in surrogate recipients, such as a cow or sheep in the case of farm animals.

SCNT is seen as a good method for producing agriculture animals for food consumption. It successfully cloned sheep, cattle, goats, and pigs. Another benefit is SCNT is seen as a solution to clone endangered species that are on the verge of going extinct. However, stresses placed on both the egg cell and the introduced nucleus can be enormous, which led to a high loss in resulting cells in early research. For example, the cloned sheep Dolly was born after 277 eggs were used for SCNT, which created 29 viable embryos. Only three of these embryos survived until birth, and only one survived to adulthood. As the procedure could not be automated, and had to be performed manually under a microscope, SCNT was very resource intensive. The biochemistry involved in reprogramming the differentiated somatic cell nucleus and activating the recipient egg was also far from being well-understood. However, by 2014 researchers were reporting cloning success rates of seven to eight out of ten and in 2016, a Korean Company Sooam Biotech was reported to be producing 500 cloned embryos per day.

In SCNT, not all of the donor cell's genetic information is transferred, as the donor cell's mitochondria that contain their own mitochondrial DNA are left behind. The resulting hybrid cells retain those mitochondrial structures which originally belonged to the egg. As a consequence, clones such as Dolly that are born from SCNT are not perfect copies of the donor of the nucleus.

Organism Cloning

Organism cloning (also called reproductive cloning) refers to the procedure of creating a new multicellular organism, genetically identical to another. In essence this form of cloning is an asexual method of reproduction, where fertilization or inter-gamete contact does not take place. Asexual reproduction is a naturally occurring phenomenon in many species, including most plants and some insects. Scientists have made some major achievements with cloning, including the asexual reproduction of sheep and cows. There is a lot of ethical debate over whether or not cloning should be used. However, cloning, or asexual propagation, has been common practice in the horticultural world for hundreds of years.

Horticultural

The term *clone* is used in horticulture to refer to descendants of a single plant which were produced by vegetative reproduction or apomixis. Many horticultural plant cultivars are clones, having been derived from a single individual, multiplied by some process other than sexual reproduction. As an example, some European cultivars of grapes represent clones that have been propagated for over two millennia. Other examples are potato and banana. Grafting can be regarded as cloning, since all the shoots and branches coming from the graft are genetically a clone of a single individual, but this particular kind of cloning has not come under ethical scrutiny and is generally treated as an entirely different kind of operation.

Many trees, shrubs, vines, ferns and other herbaceous perennials form clonal colonies naturally. Parts of an individual plant may become detached by fragmentation and grow on to become separate clonal individuals. A common example is in the vegetative reproduction of moss and liverwort gametophyte clones by means of gemmae. Some vascular plants e.g. dandelion and certain viviparous grasses also form seeds asexually, termed apomixis, resulting in clonal populations of genetically identical individuals.

Parthenogenesis

Clonal derivation exists in nature in some animal species and is referred to as parthenogenesis (reproduction of an organism by itself without a mate). This is an asexual form of reproduction that is only found in females of some insects, crustaceans, nematodes, fish (for example the hammerhead shark), the Komodo dragon and lizards. The growth and development occurs without fertilization by a male. In plants, parthenogenesis means the development of an embryo from an unfertilized egg cell, and is a component process of apomixis. In species that use the XY sex-determination system, the offspring will always be female. An example is the little fire ant (*Wasmannia auropunctata*), which is native to Central and South America but has spread throughout many tropical environments.

Artificial Cloning of Organisms

Artificial cloning of organisms may also be called *reproductive cloning*.

First Moves

Hans Spemann, a German embryologist was awarded a Nobel Prize in Physiology or Medicine in 1935 for his discovery of the effect now known as embryonic induction, exercised by various parts of the embryo, that directs the development of groups of cells into particular tissues and organs.

In 1928 he and his student, Hilde Mangold, were the first to perform somatic-cell nuclear transfer using amphibian embryos – one of the first moves towards cloning.

Methods

Reproductive cloning generally uses "somatic cell nuclear transfer" (SCNT) to create animals that are genetically identical. This process entails the transfer of a nucleus from a donor adult cell (somatic cell) to an egg from which the nucleus has been removed, or to a cell from a blastocyst from which the nucleus has been removed. If the egg begins to divide normally it is transferred into the uterus of the surrogate mother. Such clones are not strictly identical since the somatic cells may contain mutations in their nuclear DNA. Additionally, the mitochondria in the cytoplasm also contains DNA and during SCNT this mitochondrial DNA is wholly from the cytoplasmic donor's egg, thus the mitochondrial genome is not the same as that of the nucleus donor cell from which it was produced. This may have important implications for cross-species nuclear transfer in which nuclear-mitochondrial incompatibilities may lead to death.

Artificial *embryo splitting* or *embryo twinning*, a technique that creates monozygotic twins from a single embryo, is not considered in the same fashion as other methods of cloning. During that procedure, an donor embryo is split in two distinct embryos, that can then be transferred via embryo transfer. It is optimally performed at the 6- to 8-cell stage, where it can be used as an expansion of IVF to increase the number of available embryos. If both embryos are successful, it gives rise to monozygotic (identical) twins.

Dolly the Sheep

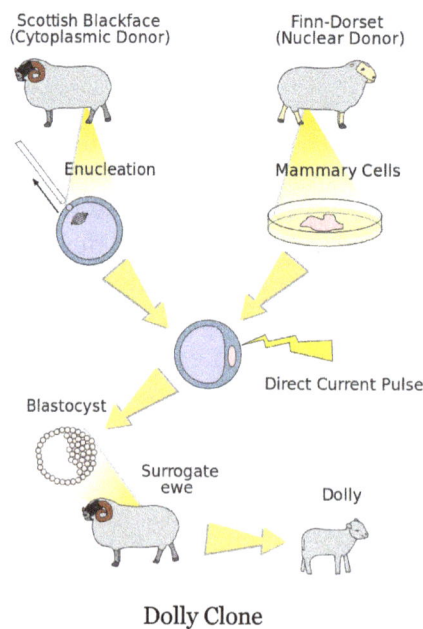

Dolly Clone

Dolly, a Finn-Dorset ewe, was the first mammal to have been successfully cloned from an adult somatic cell. Dolly was formed by taking a cell from the udder of her 6-year old biological mother. Dolly's embryo was created by taking the cell and inserting it into a sheep ovum. It took 434 attempts before an embryo was successful. The embryo was then placed inside a female sheep that went through a normal pregnancy. She was cloned at the Roslin Institute in Scotland and lived there from her birth in 1996 until her death in 2003 when she was six. She was born on 5 July 1996 but not announced to the world until 22 February 1997. Her stuffed remains were placed at Edinburgh's Royal Museum, part of the National Museums of Scotland.

Dolly was publicly significant because the effort showed that genetic material from a specific adult cell, programmed to express only a distinct subset of its genes, can be reprogrammed to grow an entirely new organism. Before this demonstration, it had been shown by John Gurdon that nuclei from differentiated cells could give rise to an entire organism after transplantation into an enucleated egg. However, this concept was not yet demonstrated in a mammalian system.

The first mammalian cloning (resulting in Dolly the sheep) had a success rate of 29 embryos per 277 fertilized eggs, which produced three lambs at birth, one of which lived. In a bovine experiment involving 70 cloned calves, one-third of the calves died young. The first successfully cloned horse, Prometea, took 814 attempts. Notably, although the first clones were frogs, no adult cloned frog has yet been produced from a somatic adult nucleus donor cell.

There were early claims that Dolly the sheep had pathologies resembling accelerated aging. Scientists speculated that Dolly's death in 2003 was related to the shortening of telomeres, DNA-protein complexes that protect the end of linear chromosomes. However, other researchers, including Ian Wilmut who led the team that successfully cloned Dolly, argue that Dolly's early death due to respiratory infection was unrelated to deficiencies with the cloning process. This idea that the nuclei have not irreversibly aged was shown in 2013 to be true for mice.

Dolly was named after performer Dolly Parton because the cells cloned to make her were from a mammary gland cell, and Parton is known for her ample cleavage.

Species Cloned

The modern cloning techniques involving nuclear transfer have been successfully performed on several species. Notable experiments include:

- Tadpole: (1952) Robert Briggs and Thomas J. King had successfully cloned northern leopard frogs: thirty-five complete embryos and twenty-seven tadpoles from one-hundred and four successful nuclear transfers.

- Carp: (1963) In China, embryologist Tong Dizhou produced the world's first cloned fish by inserting the DNA from a cell of a male carp into an egg from a female carp. He published the findings in a Chinese science journal.

- Mice: (1986) A mouse was successfully cloned from an early embryonic cell. Soviet scientists Chaylakhyan, Veprencev, Sviridova, and Nikitin had the mouse "Masha" cloned. Research was published in the magazine "Biofizika" volume XXXII, issue 5 of 1987.

- Sheep: Marked the first mammal being cloned (1984) from early embryonic cells by Steen Willadsen. Megan and Morag cloned from differentiated embryonic cells in June 1995 and Dolly the sheep from a somatic cell in 1996.

- Rhesus monkey: Tetra (January 2000) from embryo splitting

- Pig: the first cloned pigs (March 2000). By 2014, BGI in China was producing 500 cloned pigs a year to test new medicines.

- Gaur: (2001) was the first endangered species cloned.

- Cattle: Alpha and Beta (males, 2001) and (2005) Brazil

- Cat: CopyCat "CC" (female, late 2001), Little Nicky, 2004, was the first cat cloned for commercial reasons

- Rat: Ralph, the first cloned rat (2003)

- Mule: Idaho Gem, a john mule born 4 May 2003, was the first horse-family clone.

- Horse: Prometea, a Haflinger female born 28 May 2003, was the first horse clone.

- Dog: Snuppy, a male Afghan hound was the first cloned dog (2005).

- Wolf: Snuwolf and Snuwolffy, the first two cloned female wolves (2005).

- Water buffalo: Samrupa was the first cloned water buffalo. It was born on 6 February 2009, at India's Karnal National Diary Research Institute but died five days later due to lung infection.

- Pyrenean ibex (2009) was the first extinct animal to be cloned back to life; the clone lived for seven minutes before dying of lung defects.

- Camel: (2009) Injaz, is the first cloned camel.

- Pashmina goat: (2012) Noori, is the first cloned pashmina goat. Scientists at the faculty of veterinary sciences and animal husbandry of Sher-e-Kashmir University of Agricultural Sciences and Technology of Kashmir successfully cloned the first Pashmina goat (Noori) using the advanced reproductive techniques under the leadership of Riaz Ahmad Shah.

- Gastric brooding frog: (2013) The gastric brooding frog, *Rheobatrachus silus*, thought to have been extinct since 1983 was cloned in Australia, although the embryos died after a few days.

Human Cloning

Human cloning is the creation of a genetically identical copy of a human. The term is generally used to refer to artificial human cloning, which is the reproduction of human cells and tissues. It does not refer to the natural conception and delivery of identical twins. The possibility of human cloning has raised controversies. These ethical concerns have prompted several nations to pass legislature regarding human cloning and its legality.

Two commonly discussed types of theoretical human cloning are *therapeutic cloning* and *reproductive cloning*. Therapeutic cloning would involve cloning cells from a human for use in medicine and transplants, and is an active area of research, but is not in medical practice anywhere in the world, as of 2014. Two common methods of therapeutic cloning that are being researched are somatic-cell nuclear transfer and, more recently, pluripotent stem cell induction. Reproductive cloning would involve making an entire cloned human, instead of just specific cells or tissues.

Ethical Issues of Cloning

There are a variety of ethical positions regarding the possibilities of cloning, especially human cloning. While many of these views are religious in origin, the questions raised by cloning are faced by secular perspectives as well. Perspectives on human cloning are theoretical, as human therapeutic and reproductive cloning are not commercially used; animals are currently cloned in laboratories and in livestock production.

Advocates support development of therapeutic cloning in order to generate tissues and whole organs to treat patients who otherwise cannot obtain transplants, to avoid the need for immunosuppressive drugs, and to stave off the effects of aging. Advocates for reproductive cloning believe that parents who cannot otherwise procreate should have access to the technology.

Opponents of cloning have concerns that technology is not yet developed enough to be safe and that it could be prone to abuse (leading to the generation of humans from whom organs and tissues would be harvested), as well as concerns about how cloned individuals could integrate with families and with society at large.

Religious groups are divided, with some opposing the technology as usurping "God's place" and, to the extent embryos are used, destroying a human life; others support therapeutic cloning's potential life-saving benefits.

Cloning of animals is opposed by animal-groups due to the number of cloned animals that suffer from malformations before they die, and while food from cloned animals has been approved by the US FDA, its use is opposed by groups concerned about food safety.

Cloning Extinct and Endangered Species

Cloning, or more precisely, the reconstruction of functional DNA from extinct species has, for decades, been a dream. Possible implications of this were dramatized in the 1984 novel *Carnosaur* and the 1990 novel *Jurassic Park*. The best current cloning techniques have an average success rate of 9.4 percent (and as high as 25 percent) when working with familiar species such as mice, while cloning wild animals is usually less than 1 percent successful. Several tissue banks have come into existence, including the "Frozen Zoo" at the San Diego Zoo, to store frozen tissue from the world's rarest and most endangered species.

In 2001, a cow named Bessie gave birth to a cloned Asian gaur, an endangered species, but the calf died after two days. In 2003, a banteng was successfully cloned, followed by three African wildcats from a thawed frozen embryo. These successes provided hope that similar techniques (using surrogate mothers of another species) might be used to clone extinct species. Anticipating this possibility, tissue samples from the last *bucardo* (Pyrenean ibex) were frozen in liquid nitrogen immediately after it died in 2000. Researchers are also considering cloning endangered species such as the giant panda and cheetah.

In 2002, geneticists at the Australian Museum announced that they had replicated DNA of the thylacine (Tasmanian tiger), at the time extinct for about 65 years, using polymerase chain reaction. However, on 15 February 2005 the museum announced that it was stopping the project after tests showed the specimens' DNA had been too badly degraded by the (ethanol) preservative. On 15 May

2005 it was announced that the thylacine project would be revived, with new participation from researchers in New South Wales and Victoria.

In January 2009, for the first time, an extinct animal, the Pyrenean ibex mentioned above was cloned, at the Centre of Food Technology and Research of Aragon, using the preserved frozen cell nucleus of the skin samples from 2001 and domestic goat egg-cells. The ibex died shortly after birth due to physical defects in its lungs.

One of the most anticipated targets for cloning was once the woolly mammoth, but attempts to extract DNA from frozen mammoths have been unsuccessful, though a joint Russo-Japanese team is currently working toward this goal. In January 2011, it was reported by Yomiuri Shimbun that a team of scientists headed by Akira Iritani of Kyoto University had built upon research by Dr. Wakayama, saying that they will extract DNA from a mammoth carcass that had been preserved in a Russian laboratory and insert it into the egg cells of an African elephant in hopes of producing a mammoth embryo. The researchers said they hoped to produce a baby mammoth within six years. It was noted, however that the result, if possible, would be an elephant-mammoth hybrid rather than a true mammoth. Another problem is the survival of the reconstructed mammoth: ruminants rely on a symbiosis with specific microbiota in their stomachs for digestion.

Scientists at the University of Newcastle and University of New South Wales announced in March 2013 that the very recently extinct gastric-brooding frog would be the subject of a cloning attempt to resurrect the species.

Many such "de-extinction" projects are described in the Long Now Foundation's Revive and Restore Project.

Lifespan

After an eight-year project involving the use of a pioneering cloning technique, Japanese researchers created 25 generations of healthy cloned mice with normal lifespans, demonstrating that clones are not intrinsically shorter-lived than naturally born animals.

In a detailed study released in 2016 and less detailed studies by others suggest that once cloned animals get past the first month or two of life they are generally healthy. However, early pregnancy loss and neonatal losses are still greater with cloning than natural conception or assisted reproduction (IVF). Current research endeavors are attempting to overcome this problem.

In Popular Culture

In an article in the 8 November 1993 article of *Time*, cloning was portrayed in a negative way, modifying Michelangelo's *Creation of Adam* to depict Adam with five identical hands. *Newsweek's* 10 March 1997 issue also critiqued the ethics of human cloning, and included a graphic depicting identical babies in beakers.

Cloning is a recurring theme in a wide variety of contemporary science fiction, ranging from action films such as *Jurassic Park* (1993), *The 6th Day* (2000), *Resident Evil* (2002), *Star Wars* (2002) and *The Island* (2005), to comedies such as Woody Allen's 1973 film *Sleeper*.

Science fiction has used cloning, most commonly and specifically human cloning, due to the fact that it brings up controversial questions of identity. *A Number* is a 2002 play by English playwright Caryl Churchill which addresses the subject of human cloning and identity, especially nature and nurture. The story, set in the near future, is structured around the conflict between a father (Salter) and his sons (Bernard 1, Bernard 2, and Michael Black) – two of whom are clones of the first one. *A Number* was adapted by Caryl Churchill for television, in a co-production between the BBC and HBO Films.

A recurring sub-theme of cloning fiction is the use of clones as a supply of organs for transplantation. The 2005 Kazuo Ishiguro novel *Never Let Me Go* and the 2010 film adaption are set in an alternate history in which cloned humans are created for the sole purpose of providing organ donations to naturally born humans, despite the fact that they are fully sentient and self-aware. The 2005 film *The Island* revolves around a similar plot, with the exception that the clones are unaware of the reason for their existence.

The use of human cloning for military purposes has also been explored in several works. *Star Wars* portrays human cloning in *Clone Wars*.

The exploitation of human clones for dangerous and undesirable work was examined in the 2009 British science fiction film *Moon*. In the futuristic novel *Cloud Atlas* and subsequent film, one of the story lines focuses on a genetically-engineered fabricant clone named Sonmi~451 who is one of millions raised in an artificial "wombtank," destined to serve from birth. She is one of thousands of clones created for manual and emotional labor; Sonmi herself works as a server in a restaurant. She later discovers that the sole source of food for clones, called 'Soap', is manufactured from the clones themselves.

Cloning has been used in fiction as a way of recreating historical figures. In the 1976 Ira Levin novel *The Boys from Brazil* and its 1978 film adaptation, Josef Mengele uses cloning to create copies of Adolf Hitler.

In 2012, a Japanese television show named "Bunshin" was created. The story's main character, Mariko, is a woman studying child welfare in Hokkaido. She grew up always doubtful about the love from her mother, who looked nothing like her and who died nine years before. One day, she finds some of her mother's belongings at a relative's house, and heads to Tokyo to seek out the truth behind her birth. She later discovered that she was a clone.

In the 2013 television show *Orphan Black*, cloning is used as a scientific study on the behavioral adaptation of the clones. In a similar vein, the book *The Double* by Nobel Prize winner José Saramago explores the emotional experience of a man who discovers that he is a clone.

Genetic Engineering

Genetic engineering, also called genetic modification, is the direct manipulation of an organism's genome using biotechnology. It is a set of technologies used to change the genetic makeup of cells, including the transfer of genes within and across species boundaries to produce improved or novel organisms. New DNA may be inserted in the host genome by first isolating and copying the genetic material of interest using molecular cloning methods to generate a DNA sequence, or by synthesizing the DNA, and then inserting this construct into the host organism. Genes may be removed, or "knocked out", using a nuclease. Gene targeting is a different technique that uses homologous recombination to change an endogenous gene, and can be used to delete a gene, remove exons, add a gene, or introduce point mutations.

An organism that is generated through genetic engineering is considered to be a genetically modified organism (GMO). The first GMOs were bacteria generated in 1973 and GM mice in 1974. Insulin-producing bacteria were commercialized in 1982 and genetically modified food has been sold since 1994. GloFish, the first GMO designed as a pet, was first sold in the United States in December 2003.

Genetic engineering techniques have been applied in numerous fields including research, agriculture, industrial biotechnology, and medicine. Enzymes used in laundry detergent and medicines such as insulin and human growth hormone are now manufactured in GM cells, experimental GM cell lines and GM animals such as mice or zebrafish are being used for research purposes, and genetically modified crops have been commercialized.

IUPAC Definition

Process of inserting new genetic information into existing cells in order to modify a specific organism for the purpose of changing its characteristics.

Definition

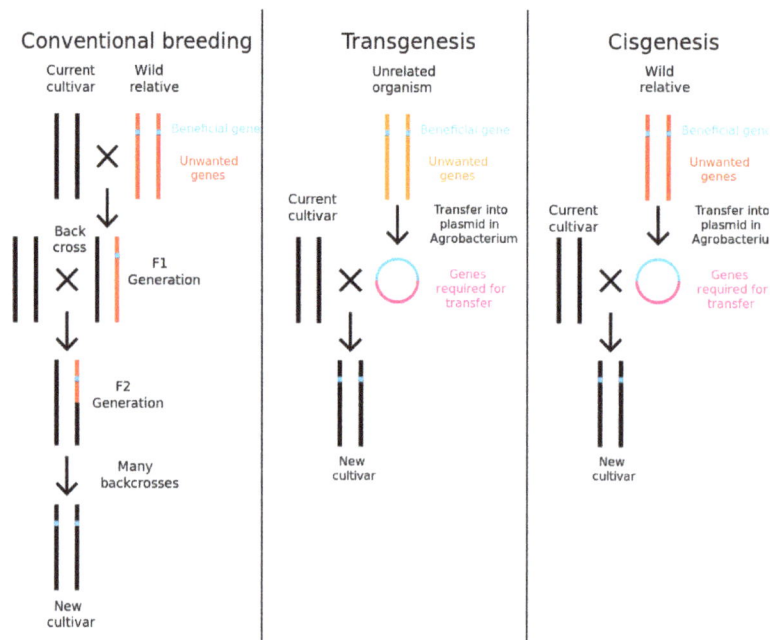

Comparison of conventional plant breeding with transgenic and cisgenic genetic modification.

Genetic engineering alters the genetic make-up of an organism using techniques that remove heritable material or that introduce DNA prepared outside the organism either directly into the host or into a cell that is then fused or hybridized with the host. This involves using recombinant nucleic acid (DNA or RNA) techniques to form new combinations of heritable genetic material followed by the incorporation of that material either indirectly through a vector system or directly through micro-injection, macro-injection and micro-encapsulation techniques.

Genetic engineering does not normally include traditional animal and plant breeding, in vitro fertilisation, induction of polyploidy, mutagenesis and cell fusion techniques that do not use recombinant nucleic acids or a genetically modified organism in the process. However the European Commission has also defined genetic engineering broadly as including selective breeding and other means of artificial selection. Cloning and stem cell research, although not considered genetic engineering, are closely related and genetic engineering can be used within them. Synthetic biology is an emerging discipline that takes genetic engineering a step further by introducing artificially synthesized material from raw materials into an organism.

If genetic material from another species is added to the host, the resulting organism is called transgenic. If genetic material from the same species or a species that can naturally breed with the host is used the resulting organism is called cisgenic. Genetic engineering can also be used to remove genetic material from the target organism, creating a gene knockout organism. In Europe genetic modification is synonymous with genetic engineering while within the United States of America it can also refer to conventional breeding methods. The Canadian regulatory system is based on whether a product has novel features regardless of method of origin. In other words, a product is regulated as genetically modified if it carries some trait not previously found in the species whether it was generated using traditional breeding methods (e.g., selective breeding, cell fusion, mutation breeding) or genetic engineering. Within the scientific community, the term *genetic engineering* is not commonly used; more specific terms such as *transgenic* are preferred.

Genetically Modified Organisms

Plants, animals or micro organisms that have changed through genetic engineering are termed genetically modified organisms or GMOs. Bacteria were the first organisms to be genetically modified. Plasmid DNA containing new genes can be inserted into the bacterial cell and the bacteria will then express those genes. These genes can code for medicines or enzymes that process food and other substrates. Plants have been modified for insect protection, herbicide resistance, virus resistance, enhanced nutrition, tolerance to environmental pressures and the production of edible vaccines. Most commercialised GMO's are insect resistant and/or herbicide tolerant crop plants. Genetically modified animals have been used for research, model animals and the production of agricultural or pharmaceutical products.

The genetically modified animals include animals with genes knocked out, increased susceptibility to disease, hormones for extra growth and the ability to express proteins in their milk.

History

Humans have altered the genomes of species for thousands of years through selective breeding, or artificial selection as contrasted with natural selection, and more recently through mutagenesis. Genetic engineering as the direct manipulation of DNA by humans outside breeding and mutations has only existed since the 1970s. The term "genetic engineering" was first coined by Jack Williamson in his science fiction novel *Dragon's Island*, published in 1951 – one year before DNA's role in heredity was confirmed by Alfred Hershey and Martha Chase, and two years before James

Watson and Francis Crick showed that the DNA molecule has a double-helix structure – though the general concept of direct genetic manipulation was explored in rudimentary form in Stanley G. Weinbaum's 1936 science fiction story *Proteus Island*.

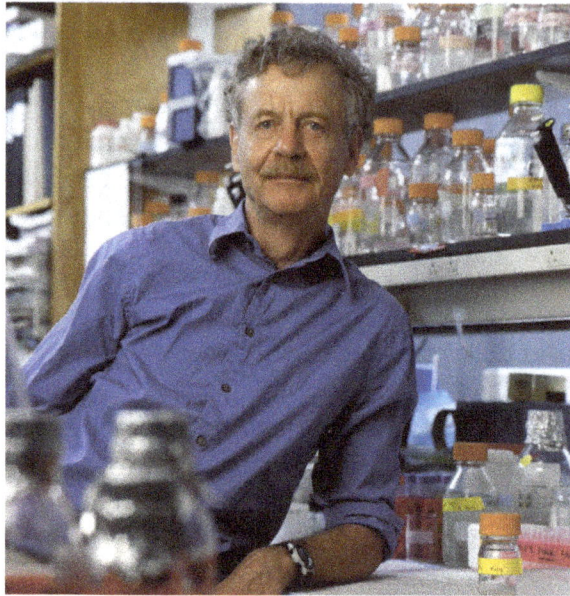

In 1974 Rudolf Jaenisch created the first GM animal.

In 1972, Paul Berg created the first recombinant DNA molecules by combining DNA from the monkey virus SV40 with that of the lambda virus. In 1973 Herbert Boyer and Stanley Cohen created the first transgenic organism by inserting antibiotic resistance genes into the plasmid of an *E. coli* bacterium. A year later Rudolf Jaenisch created a transgenic mouse by introducing foreign DNA into its embryo, making it the world's first transgenic animal. These achievements led to concerns in the scientific community about potential risks from genetic engineering, which were first discussed in depth at the Asilomar Conference in 1975. One of the main recommendations from this meeting was that government oversight of recombinant DNA research should be established until the technology was deemed safe.

In 1976 Genentech, the first genetic engineering company, was founded by Herbert Boyer and Robert Swanson and a year later the company produced a human protein (somatostatin) in *E.coli*. Genentech announced the production of genetically engineered human insulin in 1978. In 1980, the U.S. Supreme Court in the *Diamond v. Chakrabarty* case ruled that genetically altered life could be patented. The insulin produced by bacteria, branded humulin, was approved for release by the Food and Drug Administration in 1982.

In the 1970s graduate student Steven Lindow of the University of Wisconsin–Madison with D.C. Arny and C. Upper found a bacterium he identified as *P. syringae* that played a role in ice nucleation, and in 1977 he discovered a mutant ice-minus strain. Dr. Lindow (who is now a plant pathologist at the University of California-Berkeley) later successfully created a recombinant ice-minus strain. In 1983, a biotech company, Advanced Genetic Sciences (AGS) applied for U.S. government authorization to perform field tests with the ice-minus strain of *P. syringae* to protect crops from frost, but environmental groups and protestors delayed the field tests for four years with legal challenges. In 1987, the ice-minus strain of *P. syringae* became the first genetically modified organism

(GMO) to be released into the environment when a strawberry field and a potato field in California were sprayed with it. Both test fields were attacked by activist groups the night before the tests occurred: "The world's first trial site attracted the world's first field trasher".

The first field trials of genetically engineered plants occurred in France and the USA in 1986, tobacco plants were engineered to be resistant to herbicides. The People's Republic of China was the first country to commercialize transgenic plants, introducing a virus-resistant tobacco in 1992. In 1994 Calgene attained approval to commercially release the Flavr Savr tomato, a tomato engineered to have a longer shelf life. In 1994, the European Union approved tobacco engineered to be resistant to the herbicide bromoxynil, making it the first genetically engineered crop commercialized in Europe. In 1995, Bt Potato was approved safe by the Environmental Protection Agency, after having been approved by the FDA, making it the first pesticide producing crop to be approved in the USA. In 2009 11 transgenic crops were grown commercially in 25 countries, the largest of which by area grown were the USA, Brazil, Argentina, India, Canada, China, Paraguay and South Africa.

In 2010, scientists at the J. Craig Venter Institute created the first synthetic genome and inserted it into an empty bacterial cell. The resulting bacterium, named Synthia, could replicate and produce proteins. In 2014, a bacterium was developed that replicated a plasmid containing a unique base pair, creating the first organism engineered to use an expanded genetic alphabet.

Process

The first step is to choose and isolate the gene that will be inserted into the genetically modified organism. The gene can be isolated using restriction enzymes to cut DNA into fragments and gel electrophoresis to separate them out according to length. Polymerase chain reaction (PCR) can also be used to amplify up a gene segment, which can then be isolated through gel electrophoresis. If the chosen gene or the donor organism's genome has been well studied it may be present in a genetic library. If the DNA sequence is known, but no copies of the gene are available, it can be artificially synthesized.

The gene to be inserted into the genetically modified organism must be combined with other genetic elements in order for it to work properly. The gene can also be modified at this stage for better expression or effectiveness. As well as the gene to be inserted most constructs contain a promoter and terminator region as well as a selectable marker gene. The promoter region initiates transcription of the gene and can be used to control the location and level of gene expression, while the terminator region ends transcription. The selectable marker, which in most cases confers antibiotic resistance to the organism it is expressed in, is needed to determine which cells are transformed with the new gene. The constructs are made using recombinant DNA techniques, such as restriction digests, ligations and molecular cloning. The manipulation of the DNA generally occurs within a plasmid.

The most common form of genetic engineering involves inserting new genetic material randomly within the host genome. Other techniques allow new genetic material to be inserted at a specific location in the host genome or generate mutations at desired genomic loci capable of knocking out endogenous genes. The technique of gene targeting uses homologous recombination to target desired changes to a specific endogenous gene. This tends to occur at a relatively low frequency in plants and animals and generally requires the use of selectable markers. The frequency of gene targeting can be greatly enhanced with the use of engineered nucleases such as zinc finger nucleases, engineered homing endonucleases, or nucleases created from TAL effectors.

In addition to enhancing gene targeting, engineered nucleases can also be used to introduce mutations at endogenous genes that generate a gene knockout.

Transformation

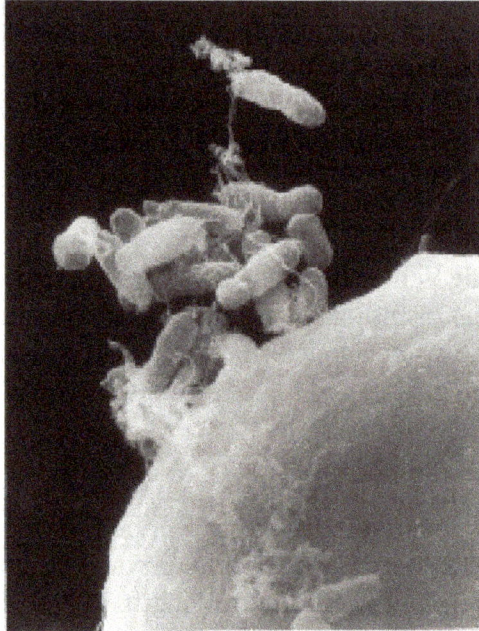

A. tumefaciens attaching itself to a carrot cell

Only about 1% of bacteria are naturally capable of taking up foreign DNA. However, this ability can be induced in other bacteria via stress (e.g. thermal or electric shock), thereby increasing the cell membrane's permeability to DNA; up-taken DNA can either integrate with the genome or exist as extrachromosomal DNA. DNA is generally inserted into animal cells using microinjection, where it can be injected through the cell's nuclear envelope directly into the nucleus or through the use of viral vectors. In plants the DNA is generally inserted using *Agrobacterium*-mediated recombination or biolistics.

In *Agrobacterium*-mediated recombination, the plasmid construct contains T-DNA, DNA which is responsible for insertion of the DNA into the host plants genome. This plasmid is transformed into *Agrobacterium* containing no plasmids prior to infecting the plant cells. The *Agrobacterium* will then naturally insert the genetic material into the plant cells. In biolistics transformation particles of gold or tungsten are coated with DNA and then shot into young plant cells or plant embryos. Some genetic material will enter the cells and transform them. This method can be used on plants that are not susceptible to *Agrobacterium* infection and also allows transformation of plant plastids. Another transformation method for plant and animal cells is electroporation. Electroporation involves subjecting the plant or animal cell to an electric shock, which can make the cell membrane permeable to plasmid DNA. In some cases the electroporated cells will incorporate the DNA into their genome. Due to the damage caused to the cells and DNA the transformation efficiency of biolistics and electroporation is lower than agrobacterial mediated transformation and microinjection.

As often only a single cell is transformed with genetic material the organism must be regenerated from that single cell. As bacteria consist of a single cell and reproduce clonally regeneration is not necessary. In plants this is accomplished through the use of tissue culture. Each plant species has different requirements for successful regeneration through tissue culture. If successful an adult plant is produced that contains the transgene in every cell. In animals it is necessary to ensure that the inserted DNA is present in the embryonic stem cells. Selectable markers are used to easily differentiate transformed from untransformed cells. These markers are usually present in the transgenic organism, although a number of strategies have been developed that can remove the selectable marker from the mature transgenic plant. When the offspring is produced they can be screened for the presence of the gene. All offspring from the first generation will be heterozygous for the inserted gene and must be mated together to produce a homozygous animal.

Further testing uses PCR, Southern hybridization, and DNA sequencing is conducted to confirm that an organism contains the new gene. These tests can also confirm the chromosomal location and copy number of the inserted gene. The presence of the gene does not guarantee it will be expressed at appropriate levels in the target tissue so methods that look for and measure the gene products (RNA and protein) are also used. These include northern hybridization, quantitative RT-PCR, Western blot, immunofluorescence, ELISA and phenotypic analysis. For stable transformation the gene should be passed to the offspring in a Mendelian inheritance pattern, so the organism's offspring are also studied.

Genome Editing

Genome editing is a type of genetic engineering in which DNA is inserted, replaced, or removed from a genome using artificially engineered nucleases, or "molecular scissors." The nucleases create specific double-stranded breaks (DSBs) at desired locations in the genome, and harness the cell's endogenous mechanisms to repair the induced break by natural processes of homologous recombination (HR) and nonhomologous end-joining (NHEJ). There are currently four families of engineered nucleases: meganucleases, zinc finger nucleases (ZFNs), transcription activator-like effector nucleases (TALENs), and the Cas9-guideRNA system (adapted from the CRISPR prokarotic immune system). In contrast to artificial genome editing natural genome editing occurs through viral and sub-viral agents competent in identification of genetic syntax structures for insertion/deletion processes with the result of conserved selection processes.

Applications

Genetic engineering has applications in medicine, research, industry and agriculture and can be used on a wide range of plants, animals and micro organisms.

Medicine

In medicine, genetic engineering has been used in manufacturing drugs, to create model animals and do laboratory research, and in gene therapy.

Manufacturing

Genetic engineering is used to mass-produce insulin, human growth hormones, follistim (for treating infertility), human albumin, monoclonal antibodies, antihemophilic factors, vaccines and many other drugs. Mouse hybridomas, cells fused together to create monoclonal antibodies, have been humanised through genetic engineering to create human monoclonal antibodies. Genetically engineered viruses are being developed that can still confer immunity, but lack the infectious sequences.

Research

Genetic engineering is used to create animal models of human diseases. Genetically modified mice are the most common genetically engineered animal model. They have been used to study and model cancer (the oncomouse), obesity, heart disease, diabetes, arthritis, substance abuse, anxiety, aging and Parkinson disease. Potential cures can be tested against these mouse models. Also genetically modified pigs have been bred with the aim of increasing the success of pig to human organ transplantation.

Gene Therapy

Gene therapy is the genetic engineering of humans, generally by replacing defective genes with effective ones. This can occur in somatic tissue or germline tissue.

Somatic gene therapy has been studied in clinical research in several diseases, including X-linked SCID, chronic lymphocytic leukemia (CLL), and Parkinson's disease. In 2012, Glybera became the first gene therapy treatment to be approved for clinical use in either Europe or the United States after its endorsement by the European Commission.

With regard to germline gene therapy, the scientific community has been opposed to attempts to alter genes in humans in inheritable ways using biotechnology since the technology was first introduced, and the caution has continued as the technology has progressed. With the advent of new techniques like CRISPR, in March 2015 scientists urged a worldwide ban on clinical use of gene editing technologies to edit the human genome in a way that can be inherited. In April 2015, Chinese researchers sparked controversy when they reported results of basic research experiments in which they edited the DNA of non-viable human embryos using CRISPR. In December 2015, scientists of major world academies called for a moratorium on inheritable human genome edits, including those related to CRISPR-Cas9 technologies.

There are also ethical concerns should the technology be used not just for treatment, but for enhancement, modification or alteration of a human beings' appearance, adaptability, intelligence, character or behavior. The distinction between cure and enhancement can also be difficult to establish. Transhumanists consider the enhancement of humans desirable.

Research

Genetic engineering is an important tool for natural scientists. Genes and other genetic information from a wide range of organisms are transformed into bacteria for storage and modification, creating genetically modified bacteria in the process. Bacteria are cheap, easy to grow, clonal, multiply quickly, relatively easy to transform and can be stored at -80 °C almost indefinitely. Once a gene is isolated it can be stored inside the bacteria providing an unlimited supply for research.

Knockout mice

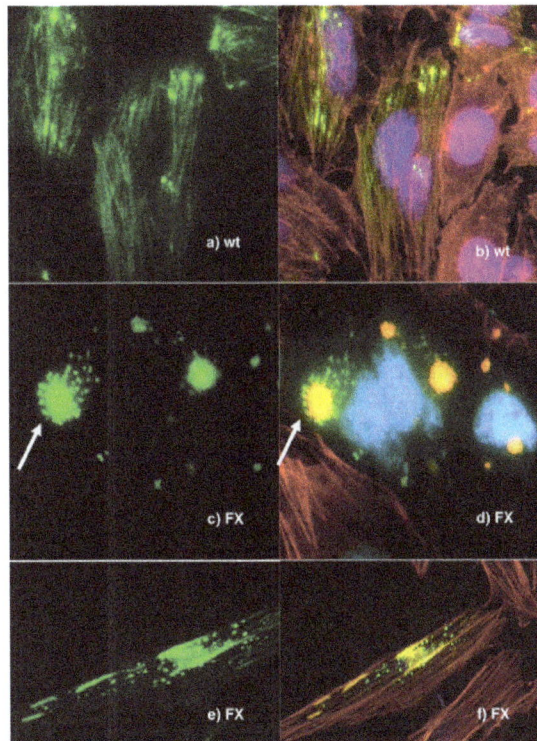

Human cells in which some proteins are fused with green fluorescent protein to allow them to be visualised

Organisms are genetically engineered to discover the functions of certain genes. This could be the effect on the phenotype of the organism, where the gene is expressed or what other genes it interacts with. These experiments generally involve loss of function, gain of function, tracking and expression.

- Loss of function experiments, such as in a gene knockout experiment, in which an organism is engineered to lack the activity of one or more genes. A knockout experiment involves the creation and manipulation of a DNA construct *in vitro*, which, in a simple knockout, consists of a copy of the desired gene, which has been altered such that it is non-functional. Embryonic stem cells incorporate the altered gene, which replaces the already present functional copy. These stem cells are injected into blastocysts, which are implanted into surrogate mothers. This allows the experimenter to analyze the defects caused by this mutation and thereby determine the role of particular genes. It is used especially frequently in developmental biology. Another method, useful in organisms such as Drosophila (fruit fly), is to induce mutations in a large population and then screen the progeny for the desired mutation. A similar process can be used in both plants and prokaryotes. Loss of function tells whether or not a protein is required for a function, but it does not always mean it's sufficient, especially if a function requires multiple proteins and lose the said function if one protein is missing.

- Gain of function experiments, the logical counterpart of knockouts. These are sometimes performed in conjunction with knockout experiments to more finely establish the function of the desired gene. The process is much the same as that in knockout engineering, except that the construct is designed to increase the function of the gene, usually by providing extra copies of the gene or inducing synthesis of the protein more frequently. Gain of function is used to tell whether or not a protein is sufficient for a function, but it does not always mean it's required. Especially when dealing with genetic/functional redundancy.

- Tracking experiments, which seek to gain information about the localization and interaction of the desired protein. One way to do this is to replace the wild-type gene with a 'fusion' gene, which is a juxtaposition of the wild-type gene with a reporting element such as green fluorescent protein (GFP) that will allow easy visualization of the products of the genetic modification. While this is a useful technique, the manipulation can destroy the function of the gene, creating secondary effects and possibly calling into question the results of the experiment. More sophisticated techniques are now in development that can track protein products without mitigating their function, such as the addition of small sequences that will serve as binding motifs to monoclonal antibodies.

- Expression studies aim to discover where and when specific proteins are produced. In these experiments, the DNA sequence before the DNA that codes for a protein, known as a gene's promoter, is reintroduced into an organism with the protein coding region replaced by a reporter gene such as GFP or an enzyme that catalyzes the production of a dye. Thus the time and place where a particular protein is produced can be observed. Expression studies can be taken a step further by altering the promoter to find which pieces are crucial for the proper expression of the gene and are actually bound by transcription factor proteins; this process is known as promoter bashing.

Industrial

Using genetic engineering techniques one can transform microorganisms such as bacteria or yeast, or transform cells from multicellular organisms such as insects or mammals, with a gene coding for a useful protein, such as an enzyme, so that the transformed organism will overexpress the desired protein. One can manufacture mass quantities of the protein by growing the transformed organism in bioreactor equipment using techniques of industrial fermentation, and then purifying the protein. Some genes do not work well in bacteria, so yeast, insect cells, or mammalians cells, each a eukaryote, can also be used. These techniques are used to produce medicines such as insulin, human growth hormone, and vaccines, supplements such as tryptophan, aid in the production of food (chymosin in cheese making) and fuels. Other applications involving genetically engineered bacteria being investigated involve making the bacteria perform tasks outside their natural cycle, such as making biofuels, cleaning up oil spills, carbon and other toxic waste and detecting arsenic in drinking water. Certain genetically modified microbes can also be used in biomining and bioremediation, due to their ability to extract heavy metals from their environment and incorporate them into compounds that are more easily recoverable.

Experimental, Lab Scale Industrial Applications

In materials science, a genetically modified virus has been used in an academic lab as a scaffold for assembling a more environmentally friendly lithium-ion battery.

Bacteria have been engineered to function as sensors by expressing a fluorescent protein under certain environmental conditions.

Agriculture

Bt-toxins present in peanut leaves (bottom image) protect it from extensive damage caused by European corn borer larvae (top image).

One of the best-known and controversial applications of genetic engineering is the creation and use of genetically modified crops or genetically modified organisms, such as genetically modified fish, which are used to produce genetically modified food and materials with diverse uses. There are four main goals in generating genetically modified crops.

One goal, and the first to be realized commercially, is to provide protection from environmental threats, such as cold (in the case of Ice-minus bacteria), or pathogens, such as insects or viruses, and/or resistance to herbicides. There are also fungal and virus resistant crops developed or in development. They have been developed to make the insect and weed management of crops easier and can indirectly increase crop yield.

Another goal in generating GMOs is to modify the quality of produce by, for instance, increasing the nutritional value or providing more industrially useful qualities or quantities. The Amflora potato, for example, produces a more industrially useful blend of starches. Cows have been engineered to produce more protein in their milk to facilitate cheese production. Soybeans and canola have been genetically modified to produce more healthy oils.

Another goal consists of driving the GMO to produce materials that it does not normally make. One example is "pharming", which uses crops as bioreactors to produce vaccines, drug intermediates, or drug themselves; the useful product is purified from the harvest and then used in the standard pharmaceutical production process. Cows and goats have been engineered to express drugs and other proteins in their milk, and in 2009 the FDA approved a drug produced in goat milk.

Another goal in generating GMOs, is to directly improve yield by accelerating growth, or making the organism more hardy (for plants, by improving salt, cold or drought tolerance). Salmon have been genetically modified with growth hormones to increase their size.

The genetic engineering of agricultural crops can increase the growth rates and resistance to different diseases caused by pathogens and parasites. This is beneficial as it can greatly increase the production of food sources with the usage of fewer resources that would be required to host the world's growing populations. These modified crops would also reduce the usage of chemicals, such as fertilizers and pesticides, and therefore decrease the severity and frequency of the damages produced by these chemical pollution.

Ethical and safety concerns have been raised around the use of genetically modified food. A major safety concern relates to the human health implications of eating genetically modified food, in particular whether toxic or allergic reactions could occur. Gene flow into related non-transgenic crops, off target effects on beneficial organisms and the impact on biodiversity are important environmental issues. Ethical concerns involve religious issues, corporate control of the food supply, intellectual property rights and the level of labeling needed on genetically modified products.

Conservation

Genetic engineering has potential applications in conservation and natural areas management. For example, gene transfer through viral vectors has been proposed as a means of controlling invasive species as well as vaccinating threatened fauna from disease. Transgenic trees have also been suggested as a way to confer resistance to pathogens in wild populations. With the increasing risks of maladaptation in organisms as a result of climate change and other perturbations, facilitated adaptation through gene tweaking could be one solution to reducing extinction risks. Applications of genetic engineering in conservation are thus far mostly theoretical and have yet to be put into practice. Further experimentation will be necessary to gauge the benefits and costs of such practices.

BioArt and Entertainment

Genetic engineering is also being used to create BioArt. Some bacteria have been genetically engineered to create black and white photographs.

Genetic engineering has also been used to create novelty items such as lavender-colored carnations, blue roses, and glowing fish.

Regulation

The regulation of genetic engineering concerns the approaches taken by governments to assess and manage the risks associated with the development and release of genetically modified crops. There are differences in the regulation of GM crops between countries, with some of the most marked differences occurring between the USA and Europe. Regulation varies in a given country depending on the intended use of the products of the genetic engineering. For example, a crop not intended for food use is generally not reviewed by authorities responsible for food safety. Starting in the late 1980s, guidance on assessing the safety of genetically engineered plants and food emerged from organizations including the FAO and WHO.

Controversy

Critics have objected to use of genetic engineering per se on several grounds, including ethical concerns, ecological concerns, and economic concerns raised by the fact GM techniques and GM organisms are subject to intellectual property law. GMOs also are involved in controversies over GM food with respect to whether food produced from GM crops is safe, whether it should be labeled, and whether GM crops are needed to address the world's food needs. These controversies have led to litigation, international trade disputes, and protests, and to restrictive regulation of commercial products in some countries.

Recombinant DNA

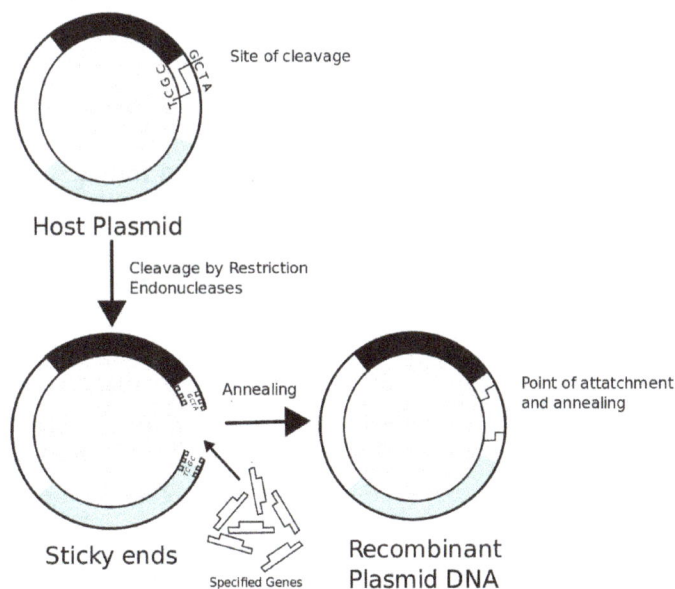

Construction of recombinant DNA, in which a foreign DNA fragment is inserted into a plasmid vector. In this example, the gene indicated by the white color is inactivated upon insertion of the foreign DNA fragment.

Recombinant DNA (rDNA) molecules are DNA molecules formed by laboratory methods of genetic recombination (such as molecular cloning) to bring together genetic material from multiple sources, creating sequences that would not otherwise be found in the genome. Recombinant DNA is possible because DNA molecules from all organisms share the same chemical structure. They-differ only in the nucleotide sequence within that identical overall structure.

Introduction

Recombinant DNA is the general name for a piece of DNA that has been created by the combination of at least two strands. Recombinant DNA molecules are sometimes called chimeric DNA, because they can be made of material from two different species, like the mythical chimera. R-DNA technology uses palindromic sequences and leads to the production of sticky and blunt ends.

The DNA sequences used in the construction of recombinant DNA molecules can originate from any species. For example, plant DNA may be joined to bacterial DNA, or human DNA may be joined with fungal DNA. In addition, DNA sequences that do not occur anywhere in nature may be created by the chemical synthesis of DNA, and incorporated into recombinant molecules. Using recombinant DNA technology and synthetic DNA, literally any DNA sequence may be created and introduced into any of a very wide range of living organisms.

Proteins that can result from the expression of recombinant DNA within living cells are termed recombinant proteins. When recombinant DNA encoding a protein is introduced into a host organism, the recombinant protein is not necessarily produced. Expression of foreign proteins requires the use of specialized expression vectors and often necessitates significant restructuring by foreign coding sequences.

Recombinant DNA differs from genetic recombination in that the former results from artificial methods in the test tube, while the latter is a normal biological process that results in the remixing of existing DNA sequences in essentially all organisms.

Creating Recombinant DNA

Molecular cloning is the laboratory process used to create recombinant DNA. It is one of two widely used methods, along with polymerase chain reaction (PCR) used to direct the replication of any specific DNA sequence chosen by the experimentalist. The fundamental difference between the two methods is that molecular cloning involves replication of the DNA within a living cell, while PCR replicates DNA in the test tube, free of living cells.

Formation of recombinant DNA requires a cloning vector, a DNA molecule that replicates within a living cell. Vectors are generally derived from plasmids or viruses, and represent relatively small segments of DNA that contain necessary genetic signals for replication, as well as additional elements for convenience in inserting foreign DNA, identifying cells that contain recombinant DNA, and, where appropriate, expressing the foreign DNA. The choice of vector for molecular cloning depends on the choice of host organism, the size of the DNA to be cloned, and whether and how the foreign DNA is to be expressed. The DNA segments can be combined by using a variety of methods, such as restriction enzyme/ligase cloning or Gibson assembly.

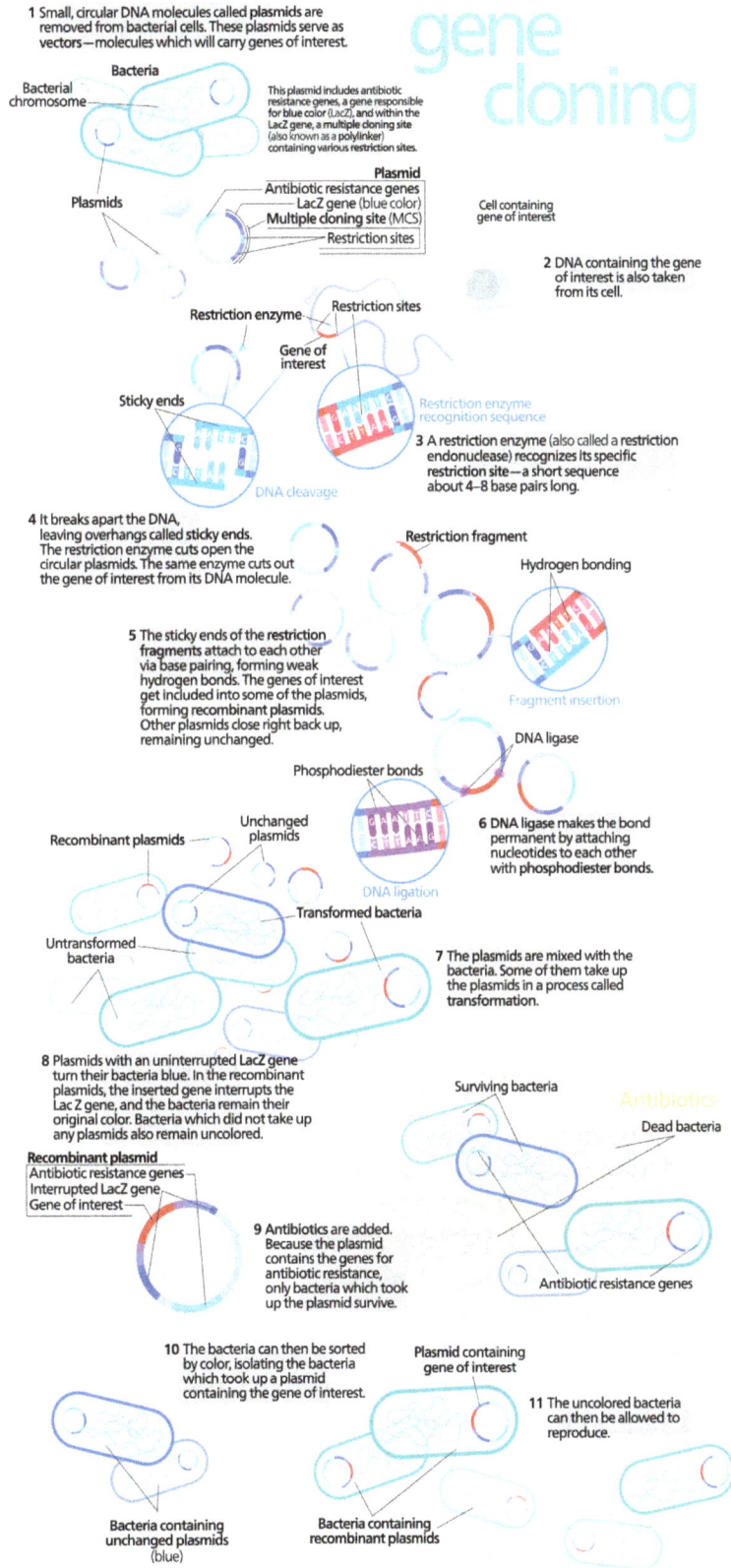

1 Small, circular DNA molecules called **plasmids** are removed from bacterial cells. These plasmids serve as vectors—molecules which will carry genes of interest.

Bacteria

Bacterial chromosome

This plasmid includes antibiotic resistance genes, a gene responsible for blue color (LacZ), and within the LacZ gene, a multiple cloning site (also known as a polylinker) containing various restriction sites.

Plasmids

Plasmid
Antibiotic resistance genes
LacZ gene (blue color)
Multiple cloning site (MCS)
Restriction sites

Cell containing gene of interest

2 DNA containing the gene of interest is also taken from its cell.

Restriction enzyme — Restriction sites

Gene of interest

Sticky ends

Restriction enzyme recognition sequence

3 A **restriction enzyme** (also called a restriction endonuclease) recognizes its specific restriction site—a short sequence about 4–8 base pairs long.

DNA cleavage

4 It breaks apart the DNA, leaving overhangs called **sticky ends**. The restriction enzyme cuts open the circular plasmids. The same enzyme cuts out the gene of interest from its DNA molecule.

Restriction fragment

Hydrogen bonding

5 The sticky ends of the **restriction fragments** attach to each other via base pairing, forming weak hydrogen bonds. The genes of interest get included into some of the plasmids, forming recombinant plasmids. Other plasmids close right back up, remaining unchanged.

Fragment insertion

DNA ligase

Phosphodiester bonds

Unchanged plasmids

Recombinant plasmids

6 **DNA ligase** makes the bond permanent by attaching nucleotides to each other with **phosphodiester bonds**.

DNA ligation

Transformed bacteria

Untransformed bacteria

7 The plasmids are mixed with the bacteria. Some of them take up the plasmids in a process called transformation.

8 Plasmids with an uninterrupted LacZ gene turn their bacteria blue. In the recombinant plasmids, the inserted gene interrupts the Lac Z gene, and the bacteria remain their original color. Bacteria which did not take up any plasmids also remain uncolored.

Surviving bacteria

Antibiotics

Dead bacteria

Recombinant plasmid
Antibiotic resistance genes
Interrupted LacZ gene
Gene of interest

9 Antibiotics are added. Because the plasmid contains the genes for antibiotic resistance, only bacteria which took up the plasmid survive.

Antibiotic resistance genes

10 The bacteria can then be sorted by color, isolating the bacteria which took up a plasmid containing the gene of interest.

Plasmid containing gene of interest

11 The uncolored bacteria can then be allowed to reproduce.

Bacteria containing unchanged plasmids (blue)

Bacteria containing recombinant plasmids

gene cloning

In standard cloning protocols, the cloning of any DNA fragment essentially involves seven steps:

(1) Choice of host organism and cloning vector, (2) Preparation of vector DNA, (3) Preparation of DNA to be cloned, (4) Creation of recombinant DNA, (5) Introduction of recombinant DNA into the host organism, (6) Selection of organisms containing recombinant DNA, and (7) Screening for clones with desired DNA inserts and biological properties.

Expression of Recombinant DNA

Following transplantation into the host organism, the foreign DNA contained within the recombinant DNA construct may or may not be expressed. That is, the DNA may simply be replicated without expression, or it may be transcribed and translated at a recombinant protein is produced. Generally speaking, expression of a foreign gene requires restructuring the gene to include sequences that are required for producing an mRNA molecule that can be used by the host's translational apparatus (e.g. promoter, translational initiation signal, and transcriptional terminator). Specific changes to the host organism may be made to improve expression of the ectopic gene. In addition, changes may be needed to the coding sequences as well, to optimize translation, make the protein soluble, direct the recombinant protein to the proper cellular or extracellular location, and stabilize the protein from degradation.

Properties of Organisms Containing Recombinant DNA

In most cases, organisms containing recombinant DNA have apparently normal phenotypes. That is, their appearance, behavior and metabolism are usually unchanged, and the only way to demonstrate the presence of recombinant sequences is to examine the DNA itself, typically using a polymerase chain reaction (PCR) test. Significant exceptions exist, and are discussed below.

If the rDNA sequences encode a gene that is expressed, then the presence of RNA and/or protein products of the recombinant gene can be detected, typically using RT-PCR or western hybridization methods. Gross phenotypic changes are not the norm, unless the recombinant gene has been chosen and modified so as to generate biological activity in the host organism. Additional phenotypes that are encountered include toxicity to the host organism induced by the recombinant gene product, especially if it is over-expressed or expressed within inappropriate cells or tissues.

In some cases, recombinant DNA can have deleterious effects even if it is not expressed. One mechanism by which this happens is insertional inactivation, in which the rDNA becomes inserted into a host cell's gene. In some cases, researchers use this phenomenon to "knock out" genes to determine their biological function and importance. Another mechanism by which rDNA insertion into chromosomal DNA can affect gene expression is by inappropriate activation of previously unexpressed host cell genes. This can happen, for example, when a recombinant DNA fragment containing an active promoter becomes located next to a previously silent host cell gene, or when a host cell gene that functions to restrain gene expression undergoes insertional inactivation by recombinant DNA.

Applications of Recombinant DNA Technology

Recombinant DNA is widely used in biotechnology, medicine and research. Today, recombinant proteins and other products that result from the use of rDNA technology are found in essentially every western pharmacy, doctor's or veterinarian's office, medical testing laboratory, and biological research laboratory. In addition, organisms that have been manipulated using recombinant DNA technology, as well as products derived from those organisms, have found their way into many farms, supermarkets, home medicine cabinets, and even pet shops, such as those that sell GloFish and other genetically modified animals.

A group of GloFish fluorescent fish

The most common application of recombinant DNA is in basic research, in which the technology is important to most current work in the biological and biomedical sciences. Recombinant DNA is used to identify, map and sequence genes, and to determine their function. rDNA probes are employed in analyzing gene expression within individual cells, and throughout the tissues of whole organisms. Recombinant proteins are widely used as reagents in laboratory experiments and to generate antibody probes for examining protein synthesis within cells and organisms.

Many additional practical applications of recombinant DNA are found in industry, food production, human and veterinary medicine, agriculture, and bioengineering. Some specific examples are identified below.

Recombinant Chymosin

Found in rennet, chymosin is an enzyme required to manufacture cheese. It was the first genetically engineered food additive used commercially. Traditionally, processors obtained chymosin from rennet, a preparation derived from the fourth stomach of milk-fed calves. Scientists engineered a non-pathogenic strain (K-12) of *E. coli* bacteria for large-scale laboratory production of the enzyme. This microbiologically produced recombinant enzyme, identical structurally to the calf derived enzyme, costs less and is produced in abundant quantities. Today about 60% of U.S. hard cheese is made with genetically engineered chymosin. In 1990, FDA granted chymosin "generally recognized as safe" (GRAS) status based on data showing that the enzyme was safe.

Recombinant Human Insulin

Almost completely replaced insulin obtained from animal sources (e.g. pigs and cattle) for the treatment of insulin-dependent diabetes. A variety of different recombinant insulin preparations are in widespread use. Recombinant insulin is synthesized by inserting the human insulin gene into *E. coli*, or yeast (saccharomyces cerevisiae) which then produces insulin for human use.

Recombinant Human Growth Hormone (HGH, Somatotropin)

Administered to patients whose pituitary glands generate insufficient quantities to support normal growth and development. Before recombinant HGH became available, HGH for therapeutic use was obtained from pituitary glands of cadavers. This unsafe practice led to some patients developing Creutzfeldt–Jakob disease. Recombinant HGH eliminated this problem, and is now used therapeutically. It has also been misused as a performance-enhancing drug by athletes and others. DrugBank entry

Recombinant Blood Clotting Factor VIII

A blood-clotting protein that is administered to patients with forms of the bleeding disorder hemophilia, who are unable to produce factor VIII in quantities sufficient to support normal blood coagulation. Before the development of recombinant factor VIII, the protein was obtained by processing large quantities of human blood from multiple donors, which carried a very high risk of transmission of blood borne infectious diseases, for example HIV and hepatitis B. DrugBank entry

Recombinant Hepatitis B Vaccine

Hepatitis B infection is controlled through the use of a recombinant hepatitis B vaccine, which contains a form of the hepatitis B virus surface antigen that is produced in yeast cells. The development of the recombinant subunit vaccine was an important and necessary development because hepatitis B virus, unlike other common viruses such as polio virus, cannot be grown in vitro. Vaccine information from Hepatitis B Foundation

Diagnosis of Infection with HIV

Each of the three widely used methods for diagnosing HIV infection has been developed using recombinant DNA. The antibody test (ELISA or western blot) uses a recombinant HIV protein to test for the presence of antibodies that the body has produced in response to an HIV infection. The DNA test looks for the presence of HIV genetic material using reverse transcription polymerase chain reaction (RT-PCR). Development of the RT-PCR test was made possible by the molecular cloning and sequence analysis of HIV genomes. HIV testing page from US Centers for Disease Control (CDC)

Golden Rice

A recombinant variety of rice that has been engineered to express the enzymes responsible for β-carotene biosynthesis. This variety of rice holds substantial promise for reducing the incidence of vitamin A deficiency in the world's population. Golden rice is not currently in use, pending the resolution of regulatory and intellectual property issues.

Herbicide-resistant Crops

Commercial varieties of important agricultural crops (including soy, maize/corn, sorghum, canola, alfalfa and cotton) have been developed that incorporate a recombinant gene that results in resistance to the herbicide glyphosate (trade name *Roundup*), and simplifies weed control by glyphosate application. These crops are in common commercial use in several countries.

Insect-resistant Crops

Bacillus thuringeiensis is a bacterium that naturally produces a protein (Bt toxin) with insecticidal properties. The bacterium has been applied to crops as an insect-control strategy for many years, and this practice has been widely adopted in agriculture and gardening. Recently, plants have been developed that express a recombinant form of the bacterial protein, which may effectively control some insect predators. Environmental issues associated with the use of these transgenic crops have not been fully resolved.

History of Recombinant DNA

The idea of recombinant DNA was first proposed by Peter Lobban, a graduate student of Prof. Dale Kaiser in the Biochemistry Department at Stanford University Medical School. The first publications describing the successful production and intracellular replication of recombinant DNA appeared in 1972 and 1973. Stanford University applied for a US patent on recombinant DNA in 1974, listing the inventors as Stanley N. Cohen and Herbert W. Boyer; this patent was awarded in 1980. The first licensed drug generated using recombinant DNA technology was human insulin, developed by Genentech and Licensed by Eli Lilly and Company.

Controversy

Scientists associated with the initial development of recombinant DNA methods recognized that the potential existed for organisms containing recombinant DNA to have undesirable or dangerous properties. At the 1975 Asilomar Conference on Recombinant DNA, these concerns were discussed and a voluntary moratorium on recombinant DNA research was initiated for experiments that were considered particularly risky. This moratorium was widely observed until the National Institutes of Health (USA) developed and issued formal guidelines for rDNA work. Today, recombinant DNA molecules and recombinant proteins are usually not regarded as dangerous. However, concerns remain about some organisms that express recombinant DNA, particularly when they leave the laboratory and are introduced into the environment or food chain. These concerns are discussed in the articles on genetically modified organisms and genetically modified food controversies.

Tissue Engineering

Tissue engineering is the use of a combination of cells, engineering and materials methods, and suitable biochemical and physicochemical factors to improve or replace biological tissues. Tissue engineering involves the use of a scaffold for the formation of new viable tissue for a medical purpose. While it was once categorized as a sub-field of biomaterials, having grown in scope and importance it can be considered as a field in its own.

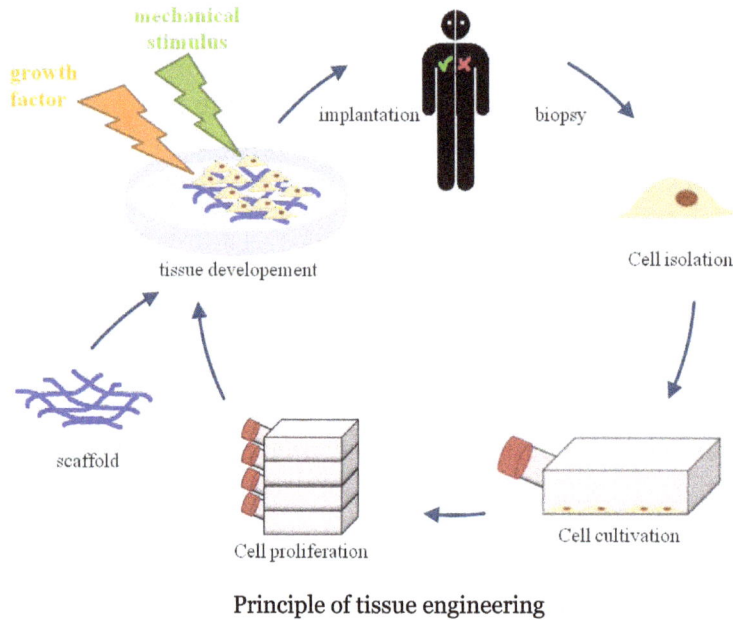

Principle of tissue engineering

While most definitions of tissue engineering cover a broad range of applications, in practice the term is closely associated with applications that repair or replace portions of or whole tissues (i.e., bone, cartilage, blood vessels, bladder, skin, muscle etc.). Often, the tissues involved require certain mechanical and structural properties for proper functioning. The term has also been applied to efforts to perform specific biochemical functions using cells within an artificially-created support system (e.g. an artificial pancreas, or a bio artificial liver). The term *regenerative medicine* is often used synonymously with tissue engineering, although those involved in regenerative medicine place more emphasis on the use of stem cells or progenitor cells to produce tissues.

Overview

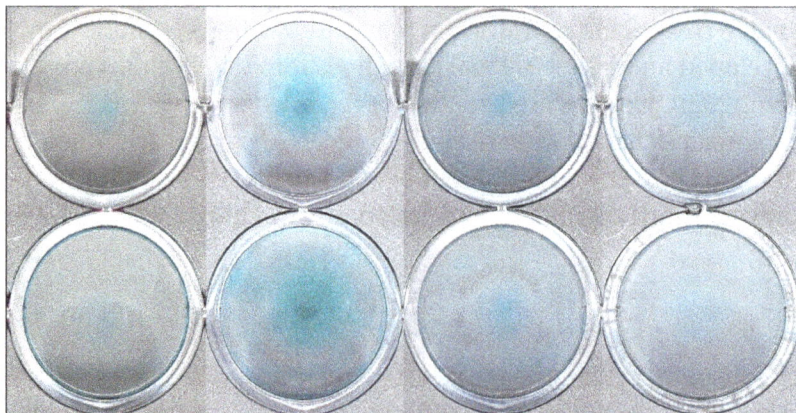

Micro-mass cultures of C3H-10T1/2 cells at varied oxygen tensions stained with Alcian blue

A commonly applied definition of tissue engineering, as stated by Langer and Vacanti, is "an interdisciplinary field that applies the principles of engineering and life sciences toward the development of biological substitutes that restore, maintain, or improve [Biological tissue] function or a whole organ". Tissue engineering has also been defined as "understanding the principles of tissue growth, and applying this to produce functional replacement tissue for clinical use." A further description goes on to say that an "underlying supposition of tissue engineering is that the employment of natural biology of the system will allow for greater success in developing therapeutic strategies aimed at the replacement, repair, maintenance, and/or enhancement of tissue function."

Powerful developments in the multidisciplinary field of tissue engineering have yielded a novel set of tissue replacement parts and implementation strategies. Scientific advances in biomaterials, stem cells, growth and differentiation factors, and biomimetic environments have created unique opportunities to fabricate tissues in the laboratory from combinations of engineered extracellular matrices ("scaffolds"), cells, and biologically active molecules. Among the major challenges now facing tissue engineering is the need for more complex functionality, as well as both functional and biomechanical stability and vascularization in laboratory-grown tissues destined for transplantation. The continued success of tissue engineering, and the eventual development of true human replacement parts, will grow from the convergence of engineering and basic research advances in tissue, matrix, growth factor, stem cell, and developmental biology, as well as materials science and bio informatics.

In 2003, the NSF published a report entitled "The Emergence of Tissue Engineering as a Research Field", which gives a thorough description of the history of this field.

Examples

- Bioartificial windpipe: The first procedure of regenerative medicine of an implantation of a "bioartificial" organ.

- In vitro meat: Edible artificial animal muscle tissue cultured *in vitro*.

- Bioartificial liver device: several research efforts have produced hepatic assist devices utilizing living hepatocytes.

- Artificial pancreas: research involves using islet cells to produce and regulate insulin, particularly in cases of diabetes.

- Artificial bladders: Anthony Atala (Wake Forest University) has successfully implanted artificially grown bladders into seven out of approximately 20 human test subjects as part of a long-term experiment.

- Cartilage: lab-grown tissue was successfully used to repair knee cartilage.

- Scaffold-free cartilage: Cartilage generated without the use of exogenous scaffold material. In this methodology, all material in the construct is cellular or material produced directly by the cells themselves.

- Doris Taylor's heart in a jar

- Tissue-engineered airway

- Tissue-engineered vessels

- Artificial skin constructed from human skin cells embedded in a hydrogel, such as in the case of bioprinted constructs for battlefield burn repairs.

- Artificial bone marrow

- Artificial bone

- Laboratory-grown penis

- Oral mucosa tissue engineering

- Foreskin

Cells as Building Blocks

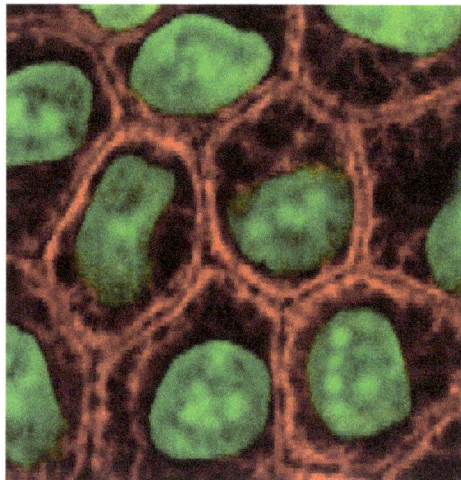

Stained cells in culture

Tissue engineering utilizes living cells as engineering materials. Examples include using living fibroblasts in skin replacement or repair, cartilage repaired with living chondrocytes, or other types of cells used in other ways.

Cells became available as engineering materials when scientists at Geron Corp. discovered how to extend telomeres in 1998, producing immortalized cell lines. Before this, laboratory cultures of healthy, noncancerous mammalian cells would only divide a fixed number of times, up to the Hayflick limit.

Extraction

From fluid tissues such as blood, cells are extracted by bulk methods, usually centrifugation or apheresis. From solid tissues, extraction is more difficult. Usually the tissue is minced, and then digested with the enzymes trypsin or collagenase to remove the extracellular matrix (ECM) that holds the cells. After that, the cells are free floating, and extracted using centrifugation or apheresis. Digestion with trypsin is very dependent on temperature. Higher temperatures digest the matrix faster, but create more damage. Collagenase is less temperature dependent, and damages fewer cells, but takes longer and is a more expensive reagent.

Types of Cells

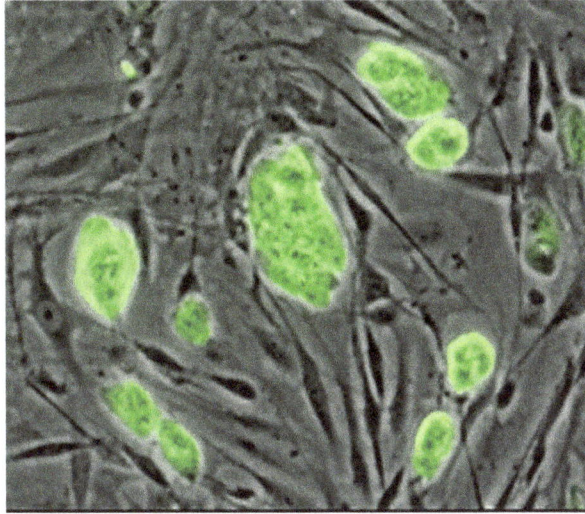

Mouse embryonic stem cells

Cells are often categorized by their source:

- *Autologous* cells are obtained from the same individual to which they will be reimplanted. Autologous cells have the fewest problems with rejection and pathogen transmission, however in some cases might not be available. For example, in genetic disease suitable autologous cells are not available. Also very ill or elderly persons, as well as patients suffering from severe burns, may not have sufficient quantities of autologous cells to establish useful cell lines. Moreover, since this category of cells needs to be harvested from the patient, there are also some concerns related to the necessity of performing such surgical operations that might lead to donor site infection or chronic pain. Autologous cells also must be cultured from samples before they can be used: this takes time, so autologous solutions may not be very quick. Recently there has been a trend towards the use of mesenchymal stem cells from bone marrow and fat. These cells can differentiate into a variety of tissue types, including bone, cartilage, fat, and nerve. A large number of cells can be easily and quickly isolated from fat, thus opening the potential for large numbers of cells to be quickly and easily obtained.

- *Allogeneic* cells come from the body of a donor of the same species. While there are some ethical constraints to the use of human cells for *in vitro* studies, the employment of dermal fibroblasts from human foreskin has been demonstrated to be immunologically safe and thus a viable choice for tissue engineering of skin.

- *Xenogenic* cells are these isolated from individuals of another species. In particular animal cells have been used quite extensively in experiments aimed at the construction of cardiovascular implants.

- *Syngenic* or *isogenic* cells are isolated from genetically identical organisms, such as twins, clones, or highly inbred research animal models.

- *Primary* cells are from an organism.

- *Secondary* cells are from a cell bank.

- *Stem cells* are undifferentiated cells with the ability to divide in culture and give rise to different forms of specialized cells. According to their source stem cells are divided into "adult" and "embryonic" stem cells, the first class being multipotent and the latter mostly pluripotent; some cells are totipotent, in the earliest stages of the embryo. While there is still a large ethical debate related with the use of embryonic stem cells, it is thought that another alternative source - induced stem cells may be useful for the repair of diseased or damaged tissues, or may be used to grow new organs.

Scaffolds

Scaffolds are materials that have been engineered to cause desirable cellular interactions to contribute to the formation of new functional tissues for medical purposes. Cells are often 'seeded' into these structures capable of supporting three-dimensional tissue formation. Scaffolds mimic the native extracellular matrix of the native tissue, recapitulating the *in vivo* milieu and allowing cells to influence their own microenvironments. They usually serve for at least one of the following purposes:

- Allow cell attachment and migration

- Deliver and retain cells and biochemical factors

- Enable diffusion of vital cell nutrients and expressed products

- Exert certain mechanical and biological influences to modify the behaviour of the cell phase

In 2009, an interdisciplinary team led by the thoracic surgeon Thorsten Walles implanted the first bioartificial transplant that provides an innate vascular network for post-transplant graft supply successfully into a patient awaiting tracheal reconstruction.

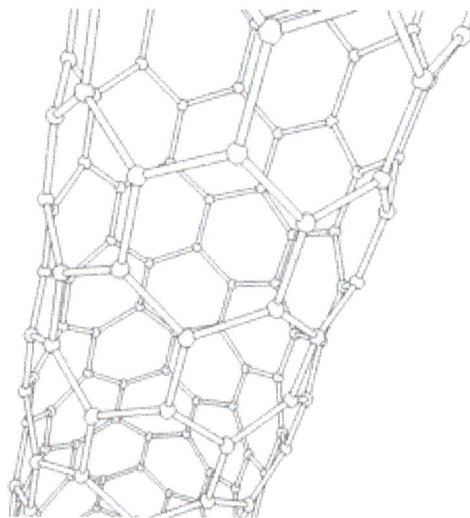

This animation of a rotating carbon nanotube shows its 3D structure. Carbon nanotubes are among the numerous candidates for tissue engineering scaffolds since they are biocompatible, resistant to biodegradation and can be functionalized with biomolecules. However, the possibility of toxicity with non-biodegradable nano-materials is not fully understood.

To achieve the goal of tissue reconstruction, scaffolds must meet some specific requirements. A high porosity and an adequate pore size are necessary to facilitate cell seeding and diffusion throughout the whole structure of both cells and nutrients. Biodegradability is often an essential factor since scaffolds should preferably be absorbed by the surrounding tissues without the necessity of a surgical removal. The rate at which degradation occurs has to coincide as much as possible with the rate of tissue formation: this means that while cells are fabricating their own natural matrix structure around themselves, the scaffold is able to provide structural integrity within the body and eventually it will break down leaving the neotissue, newly formed tissue which will take over the mechanical load. Injectability is also important for clinical uses. Recent research on organ printing is showing how crucial a good control of the 3D environment is to ensure reproducibility of experiments and offer better results.

Materials

Many different materials (natural and synthetic, biodegradable and permanent) have been investigated. Most of these materials have been known in the medical field before the advent of tissue engineering as a research topic, being already employed as bioresorbable sutures. Examples of these materials are collagen and some polyesters.

New biomaterials have been engineered to have ideal properties and functional customization: injectability, synthetic manufacture, biocompatibility, non-immunogenicity, transparency, nano-scale fibers, low concentration, resorption rates, etc. PuraMatrix, originating from the MIT labs of Zhang, Rich, Grodzinsky and Langer is one of these new biomimetic scaffold families which has now been commercialized and is impacting clinical tissue engineering.

A commonly used synthetic material is PLA - polylactic acid. This is a polyester which degrades within the human body to form lactic acid, a naturally occurring chemical which is easily removed from the body. Similar materials are polyglycolic acid (PGA) and polycaprolactone (PCL): their degradation mechanism is similar to that of PLA, but they exhibit respectively a faster and a slower rate of degradation compared to PLA. While these materials have well maintained mechanical strength and structural integrity, they exhibit a hydrophobic nature. This hydrophobicity inhibits their biocompatibility, which makes them less effective for in vivo use as tissue scaffolding. In order to fix the lack of biocompatibility, much research has been done to combine these hydrophobic materials with hydrophilic and more biocompatible hydrogels. While these hydrogels have a superior biocompatibility, they lack the structural integrity of PLA, PCL, and PGA. By combining the two different types of materials, researchers are trying to create a synergistic relationship that produces a more biocompatible tissue scaffolding. Scaffolds may also be constructed from natural materials: in particular different derivatives of the extracellular matrix have been studied to evaluate their ability to support cell growth. Proteic materials, such as collagen or fibrin, and polysaccharidic materials, like chitosan or glycosaminoglycans (GAGs), have all proved suitable in terms of cell compatibility, but some issues with potential immunogenicity still remains. Among GAGs hyaluronic acid, possibly in combination with cross linking agents (e.g. glutaraldehyde, water-soluble carbodiimide, etc.), is one of the possible choices as scaffold material. Functionalized groups of scaffolds may be useful in the delivery of small molecules (drugs) to specific tissues. Another form of scaffold under investigation is decellularised tissue extracts whereby the remaining cellular remnants/extracellular matrices act as the scaffold. Recently a range of nanocomposites bio-materials are fabricated by incorporating nanomaterials within polymeric matrix to engineer bioactive scaffolds.

A 2009 study by Ratmir et al. aimed to improve in vivo-like conditions for 3D tissue via "stacking and de-stacking layers of paper impregnated with suspensions of cells in extracellular matrix hydrogel, making it possible to control oxygen and nutrient gradients in 3D, and to analyze molecular and genetic responses". It is possible to manipulate gradients of soluble molecules, and to characterize cells in these complex gradients more effectively than conventional 3D cultures based on hydrogels, cell spheroids, or 3D perfusion reactors. Different thicknesses of paper and types of medium can support a variety of experimental environments. Upon deconstruction, these sheets can be useful in cell-based high-throughput screening and drug discovery.

Synthesis

A number of different methods have been described in literature for preparing porous structures to be employed as tissue engineering scaffolds. Each of these techniques presents its own advantages, but none are free of drawbacks.

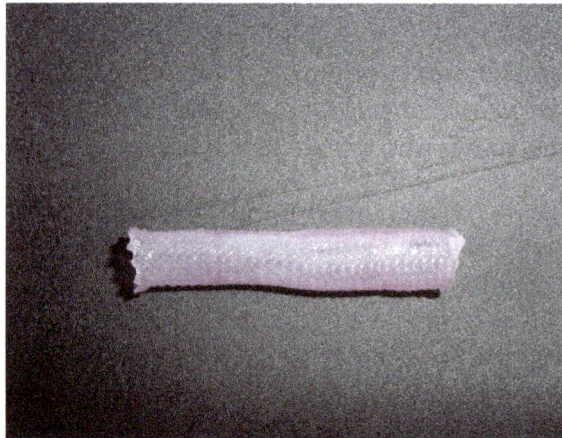

Tissue engineered vascular graft

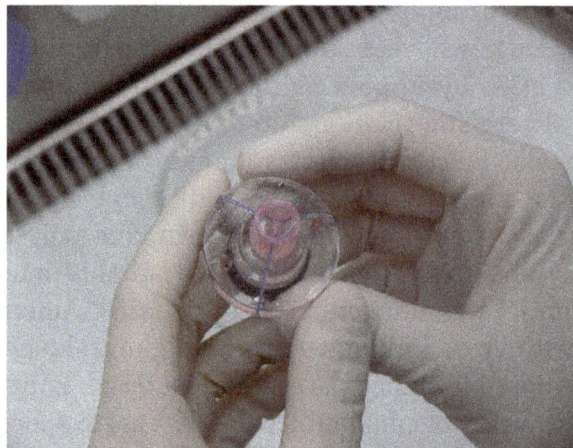

Tissue engineered heart valve

Nanofiber Self-assembly

Molecular self-assembly is one of the few methods for creating biomaterials with properties similar in scale and chemistry to that of the natural *in vivo* extracellular matrix (ECM), a crucial step toward tissue engineering of complex tissues. Moreover, these hydrogel scaffolds have shown superiority in in vivo toxicology and biocompatibility compared to traditional macroscaffolds and animal-derived materials.

Textile Technologies

These techniques include all the approaches that have been successfully employed for the preparation of non-woven meshes of different polymers. In particular, non-woven polyglycolide structures have been tested for tissue engineering applications: such fibrous structures have been found useful to grow different types of cells. The principal drawbacks are related to the difficulties in obtaining high porosity and regular pore size.

Solvent Casting and Particulate Leaching (SCPL)

This approach allows for the preparation of structures with regular porosity, but with limited thickness. First, the polymer is dissolved into a suitable organic solvent (e.g. polylactic acid could be dissolved into dichloromethane), then the solution is cast into a mold filled with porogen particles. Such porogen can be an inorganic salt like sodium chloride, crystals of saccharose, gelatin spheres or paraffin spheres. The size of the porogen particles will affect the size of the scaffold pores, while the polymer to porogen ratio is directly correlated to the amount of porosity of the final structure. After the polymer solution has been cast the solvent is allowed to fully evaporate, then the composite structure in the mold is immersed in a bath of a liquid suitable for dissolving the porogen: water in the case of sodium chloride, saccharose and gelatin or an aliphatic solvent like hexane for use with paraffin. Once the porogen has been fully dissolved, a porous structure is obtained. Other than the small thickness range that can be obtained, another drawback of SCPL lies in its use of organic solvents which must be fully removed to avoid any possible damage to the cells seeded on the scaffold.

Gas Foaming

To overcome the need to use organic solvents and solid porogens, a technique using gas as a porogen has been developed. First, disc-shaped structures made of the desired polymer are prepared by means of compression molding using a heated mold. The discs are then placed in a chamber where they are exposed to high pressure CO_2 for several days. The pressure inside the chamber is gradually restored to atmospheric levels. During this procedure the pores are formed by the carbon dioxide molecules that abandon the polymer, resulting in a sponge-like structure. The main problems resulting from such a technique are caused by the excessive heat used during compression molding (which prohibits the incorporation of any temperature labile material into the polymer matrix) and by the fact that the pores do not form an interconnected structure.

Emulsification/Freeze-drying

This technique does not require the use of a solid porogen like SCPL. First, a synthetic polymer is dissolved into a suitable solvent (e.g. polylactic acid in dichloromethane) then water is added to the polymeric solution and the two liquids are mixed in order to obtain an emulsion. Before the two phases can separate, the emulsion is cast into a mold and quickly frozen by means of immersion into liquid nitrogen. The frozen emulsion is subsequently freeze-dried to remove the dispersed water and the solvent, thus leaving a solidified, porous polymeric structure. While emulsification and freeze-drying allow for a faster preparation when compared to SCPL (since it does not require a time consuming leaching step), it still requires the use of solvents. Moreover, pore size is relatively small and porosity is often irregular. Freeze-drying by itself is also a commonly employed technique for the fabrication of scaffolds. In particular, it is used to prepare collagen sponges: collagen is dissolved into acidic solutions of acetic acid or hydrochloric acid that are cast into a mold, frozen with liquid nitrogen and then lyophilized.

Thermally Induced Phase Separation (TIPS)

Similar to the previous technique, this phase separation procedure requires the use of a solvent with a low melting point that is easy to sublime. For example, dioxane could be used to dissolve polylactic acid, then phase separation is induced through the addition of a small quantity of water: a polymer-rich and a polymer-poor phase are formed. Following cooling below the solvent melting point and some days of vacuum-drying to sublime the solvent, a porous scaffold is obtained. Liquid-liquid phase separation presents the same drawbacks of emulsification/freeze-drying.

Electrospinning

A highly versatile technique that can be used to produce continuous fibers from submicrometer to nanometer diameters. In a typical electrospinning set-up, a solution is fed through a spinneret and a high voltage is applied to the tip. The buildup of electrostatic repulsion within the charged solution, causes it to eject a thin fibrous stream. A mounted collector plate or rod with an opposite or grounded charge draws in the continuous fibers, which arrive to form a highly porous network. The primary advantages of this technique are its simplicity and ease of variation. At a laboratory level, a typical electrospinning set-up only requires a high voltage power supply (up to 30 kV), a syringe, a flat tip needle and a conducting collector. For these reasons, electrospinning has become a common method of scaffold manufacture in many labs. By modifying variables such as the distance to collector, magnitude of applied voltage, or solution flow rate—researchers can dramatically change the overall scaffold architecture.

CAD/CAM Technologies

Because most of the above techniques are limited when it comes to the control of porosity and pore size, computer assisted design and manufacturing techniques have been introduced to tissue engineering. First, a three-dimensional structure is designed using CAD software. The porosity can be tailored using algorithms within the software. The scaffold is then realized by using ink-jet printing of polymer powders or through Fused Deposition Modeling of a polymer melt.

A 2011 study by El-Ayoubi et al. investigated "3D-plotting technique to produce (biocompatible and biodegradable) poly-L-Lactide macroporous scaffolds with two different pore sizes" via solid free-form fabrication (SSF) with computer-aided-design (CAD), to explore therapeutic articular cartilage replacement as an "alternative to conventional tissue repair". The study found the smaller the pore size paired with mechanical stress in a bioreactor (to induce in vivo-like conditions), the higher the cell viability in potential therapeutic functionality via decreasing recovery time and increasing transplant effectiveness.

Laser-assisted BioPrinting (LaBP)

In a 2012 study, Koch et al. focused on whether Laser-assisted BioPrinting (LaBP) can be used to build multicellular 3D patterns in natural matrix, and whether the generated constructs are functioning and forming tissue. LaBP arranges small volumes of living cell suspensions in set high-resolution patterns. The investigation was successful, the researchers foresee that "generated tissue constructs might be used for in vivo testing by implanting them into animal models". As of this study, only human skin tissue has been synthesized, though researchers project that "by integrating further cell types (e.g. melanocytes, Schwann cells, hair follicle cells) into the printed cell construct, the behavior of these cells in a 3D in vitro microenvironment similar to their natural one can be analyzed", useful for drug discovery and toxicology studies.

Assembly Methods

One of the continuing, persistent problems with tissue engineering is mass transport limitations. Engineered tissues generally lack an initial blood supply, thus making it difficult for any implanted cells to obtain sufficient oxygen and nutrients to survive, and/or function properly.

Self-assembly

Self-assembly may play an important role here, both from the perspective of encapsulating cells and proteins, as well as creating scaffolds on the right physical scale for engineered tissue constructs and cellular ingrowth. The micromasonry is a prime technology to get cells grown in a lab to assemble into three-dimensional shapes. To break down tissue into single-cell building blocks, researchers have to dissolve the extracellular mortar that normally binds them together. But once that glue is removed, it's quite difficult to get cells to reassemble into the complex structures that make up our natural tissues. While cells aren't easily stackable, building blocks are. So the micromasonry starts with the encapsulation of living cells in polymer cubes. From there, the blocks self-assemble in any shape using templates.

Liquid-based Template Assembly

The air-liquid surface established by Faraday waves is explored as a template to assemble biological entities for bottom-up tissue engineering. This liquid-based template can be dynamically reconfigured in a few seconds, and the assembly on the template can be achieved in a scalable and parallel manner. Assembly of microscale hydrogels, cells, neuron-seeded micro-carrier beads, cell spheroids into various symmetrical and periodic structures was demonstrated with good cell viability. Formation of 3D neural network was achieved after 14-day tissue culture.

Additive Manufacturing

It might be possible to print organs, or possibly entire organisms using additive manufacturing techniques. A recent innovative method of construction uses an ink-jet mechanism to print precise layers of cells in a matrix of thermoreversible gel. Endothelial cells, the cells that line blood vessels, have been printed in a set of stacked rings. When incubated, these fused into a tube.

The field of three-dimensional and highly accurate models of biological systems is pioneered by multiple projects and technologies including a rapid method for creating tissues and even whole organs involves a 3D printer that can print the scaffolding and cells layer by layer into a working tissue sample or organ. The device is presented in a TED talk by Dr. Anthony Atala, M.D. the Director of the Wake Forest Institute for Regenerative Medicine, and the W.H. Boyce Professor and Chair of the Department of Urology at Wake Forest University, in which a kidney is printed on stage during the seminar and then presented to the crowd. It is anticipated that this technology will enable the production of livers in the future for transplantation and theoretically for toxicology and other biological studies as well.

Recently Multi-Photon Processing (MPP) was employed for in vivo expperiments by engineering artificial cartilage constructs. An ex vivo histological examination showed that certain pore geometry and the pre-growing of chondrocytes (Cho) prior to implantation significantly improves the performance of the created 3D scaffolds. The achieved biocompatibility was comparable to the commercially available collagen membranes. The successful outcome of this study supports the idea that hexagonal-pore-shaped hybrid organic-inorganic microstructured scaffolds in combination with Cho seeding may be successfully implemented for cartilage tissue engineering.

Scaffolding

In 2013, using a 3-d scaffolding of Matrigel in various configurations, substantial pancreatic organoids was produced in vitro. Clusters of small numbers of cells proliferated into 40,000 cells within one week. The clusters transform into cells that make either digestive enzymes or hormones like insulin, self-organizing into branched pancreatic organoids that resemble the pancreas.

The cells are sensitive to the environment, such as gel stiffness and contact with other cells. Individual cells do not thrive; a minimum of four proximate cells was required for subsequent organoid development. Modifications to the medium composition produced either hollow spheres mainly composed of pancreatic progenitors, or complex organoids that spontaneously undergo pancreatic morphogenesis and differentiation. Maintenance and expansion of pancreatic progenitors require active Notch and FGF signaling, recapitulating in vivo niche signaling interactions.

The organoids were seen as potentially offering mini-organs for drug testing and for spare insulin-producing cells.

Tissue Culture

In many cases, creation of functional tissues and biological structures *in vitro* requires extensive culturing to promote survival, growth and inducement of functionality. In general, the basic requirements of cells must be maintained in culture, which include oxygen, pH, humidity, temperature, nutrients and osmotic pressure maintenance.

Tissue engineered cultures also present additional problems in maintaining culture conditions. In standard cell culture, diffusion is often the sole means of nutrient and metabolite transport. However, as a culture becomes larger and more complex, such as the case with engineered organs and whole tissues, other mechanisms must be employed to maintain the culture, such as the creation of capillary networks within the tissue.

Another issue with tissue culture is introducing the proper factors or stimuli required to induce functionality. In many cases, simple maintenance culture is not sufficient. Growth factors, hormones, specific metabolites or nutrients, chemical and physical stimuli are sometimes required. For example, certain cells respond to changes in oxygen tension as part of their normal development, such as chondrocytes, which must adapt to low oxygen conditions or hypoxia during skeletal development. Others, such as endothelial cells, respond to shear stress from fluid flow, which is encountered in blood vessels. Mechanical stimuli, such as pressure pulses seem to be beneficial to all kind of cardiovascular tissue such as heart valves, blood vessels or pericardium.

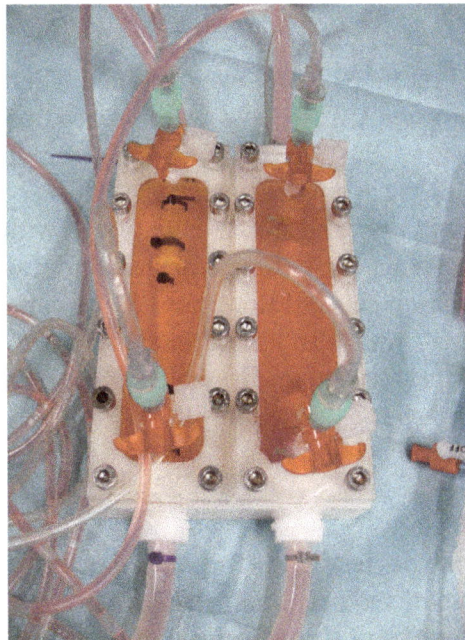

Bioreactor for cultivation of vascular grafts

Bioreactors

A bioreactor in tissue engineering, as opposed to industrial bioreactors, is a device that attempts to simulate a physiological environment in order to promote cell or tissue growth in vitro. A physiological environment can consist of many different parameters such as temperature and oxygen or carbon dioxide concentration, but can extend to all kinds of biological, chemical or mechanical stimuli. Therefore, there are systems that may include the application of forces or stresses to the tissue or even of electric current in two- or three-dimensional setups.

In academic and industry research facilities, it is typical for bioreactors to be developed to replicate the specific physiological environment of the tissue being grown (e.g., flex and fluid shearing for heart tissue growth). Several general-use and application-specific bioreactors are also commercially available, and may provide static chemical stimulation or combination of chemical and mechanical stimulation.

The Bioreactors used for 3D cell cultures are small plastic cylindrical chambers with regulated internal humidity and moisture specifically engineered for the purpose of growing cells in three dimensions. The bioreactor uses bioactive synthetic materials such as polyethylene terephthalate membranes to surround the spheroid cells in an environment that maintains high levels of nutrients. They are easy to open and close, so that cell spheroids can be removed for testing, yet the chamber is able to maintain 100% humidity throughout. This humidity is important to achieve maximum cell growth and function. The bioreactor chamber is part of a larger device that rotates to ensure equal cell growth in each direction across three dimensions. MC2 Biotek has developed a bioreactor known as ProtoTissue that uses gas exchange to maintain high oxygen levels within the cell chamber; improving upon previous bioreactors, because the higher oxygen levels help the cell grow and undergo normal cell respiration.

Long Fiber Generation

In 2013, a group from the University of Tokyo developed cell laden fibers up to a meter in length and on the order of 100 μm in size. These fibers were created using a microfluidic device that forms a double coaxial laminar flow. Each 'layer' of the microfluidic device (cells seeded in ECM, a hydrogel sheath, and finally a calcium chloride solution). The seeded cells culture within the hydrogel sheath for several days, and then the sheath is removed with viable cell fibers. Various cell types were inserted into the ECM core, including myocytes, endothelial cells, nerve cell fibers, and epithelial cell fibers. This group then showed that these fibers can be woven together to fabricate tissues or organs in a mechanism similar to textile weaving. Fibrous morphologies are advantageous in that they provide an alternative to traditional scaffold design, and many organs (such as muscle) are composed of fibrous cells.

Bioartificial Organs

An artificial organ is a man-made device that is implanted or integrated into a human to replace a natural organ, for the purpose of restoring a specific function or a group of related functions so the patient may return to a normal life as soon as possible. The replaced function doesn't necessarily have to be related to life support, but often is. The ultimate goal of tissue engineering as a discipline is to allow both 'off the shelf' bioartificial organs and regeneration of injured tissue in the body. In order to successfully create bioartificial organs from a patients stem cells, researchers continue to make improvements in the generation of complex tissues by tissue engineering. For example, much research is aimed at understanding nanoscale cues present in a cell's microenvironment.

Biopharmaceutical

A biopharmaceutical, also known as a biologic(al) medical product, biological, or biologic, is any pharmaceutical drug product manufactured in, extracted from, or semisynthesized from biological sources. Different from chemically synthesized pharmaceuticals, they include vaccines, blood, blood components, allergenics, somatic cells, gene therapies, tissues, recombinant therapeutic protein, and living cells used in cell therapy. Biologics can be composed of sugars, proteins, or nucleic acids or complex combinations of these substances, or may be living cells or tissues. They are isolated from natural sources—human, animal, or microorganism.

Terminology surrounding biopharmaceuticals varies between groups and entities, with different terms referring to different subsets of therapeutics within the general biopharmaceutical category. Some regulatory agencies use the terms *biological medicinal products* or *therapeutic biological product* to refer specifically to engineered macromolecular products like protein- and nucleic acid-based drugs, distinguishing them from products like blood, blood components, or vaccines, which are usually extracted directly from a biological source. Specialty drugs, a recent classification of pharmaceuticals, are high-cost drugs that are often biologics.

Gene-based and cellular biologics, for example, often are at the forefront of biomedical research, and may be used to treat a variety of medical conditions for which no other treatments are available.

In some jurisdictions, biologics are regulated via different pathways than other small molecule drugs and medical devices.

The term biopharmacology is sometimes used to describe the branch of pharmacology that studies biopharmaceuticals.

Major Classes

Blood plasma is a type of biopharmaceutical directly extracted from living systems.

Extracted from Living Systems

Some of the oldest forms of biologics are extracted from the bodies of animals, and other humans especially. Important biologics include:

- Whole blood and other blood components

- Organs and tissue transplants

- Stem cell therapy

- Antibodies for passive immunization (e.g., to treat a virus infection)

Some biologics that were previously extracted from animals, such as insulin, are now more commonly produced by recombinant DNA.

Produced by Recombinant DNA

As indicated the term "biologics" can be used to refer to a wide range of biological products in medicine. However, in most cases, the term "biologics" is used more restrictively for a class of therapeutics (either approved or in development) that are produced by means of biological processes involving recombinant DNA technology. These medications are usually one of three types:

1. Substances that are (nearly) identical to the body's own key signalling proteins. Examples are the blood-production stimulating protein erythropoetin, or the growth-stimulating hormone named (simply) "growth hormone" or biosynthetic human insulin and its analogues.

2. Monoclonal antibodies. These are similar to the antibodies that the human immune system uses to fight off bacteria and viruses, but they are "custom-designed" (using hybridoma technology or other methods) and can therefore be made specifically to counteract or block any given substance in the body, or to target any specific cell type; examples of such monoclonal antibodies for use in various diseases are given in the table below.

3. Receptor constructs (fusion proteins), usually based on a naturally-occurring receptor linked to the immunoglobulin frame. In this case, the receptor provides the construct with detailed specificity, whereas the immunoglobulin-structure imparts stability and other useful features in terms of pharmacology. Some examples are listed in the table below.

Biologics as a class of medications in this narrower sense have had a profound impact on many medical fields, primarily rheumatology and oncology, but also cardiology, dermatology, gastroenterology, neurology, and others. In most of these disciplines, biologics have added major therapeutic options for the treatment of many diseases, including some for which no effective therapies were available, and others where previously existing therapies were clearly inadequate. However, the advent of biologic therapeutics has also raised complex regulatory issues, and significant pharmacoeconomic concerns, because the cost for biologic therapies has been dramatically higher than for conventional (pharmacological) medications. This factor has been particularly relevant since many biological medications are used for the treatment of chronic diseases, such as rheumatoid arthritis or inflammatory bowel disease, or for the treatment of otherwise untreatable cancer during the remainder of life. The cost of treatment with a typical monoclonal antibody therapy for relatively common indications is generally in the range of €7,000–14,000 per patient per year.

Older patients who receive biologic therapy for diseases such as rheumatoid arthritis, psoriatic arthritis, or ankylosing spondylitis are at increased risk for life-threatening infection, adverse cardiovascular events, and malignancy.

The first such substance approved for therapeutic use was biosynthetic "human" insulin made via recombinant DNA. Sometimes referred to as rHI, under the trade name Humulin, was developed by Genentech, but licensed to Eli Lilly and Company, who manufactured and marketed it starting in 1982.

Major kinds of biopharmaceuticals include:

- Blood factors (Factor VIII and Factor IX)
- Thrombolytic agents (tissue plasminogen activator)
- Hormones (insulin, glucagon, growth hormone, gonadotrophins)
- Haematopoietic growth factors (Erythropoietin, colony stimulating factors)
- Interferons (Interferons-α, -β, -γ)
- Interleukin-based products (Interleukin-2)
- Vaccines (Hepatitis B surface antigen)
- Monoclonal antibodies (Various)
- Additional products (tumour necrosis factor, therapeutic enzymes)

Research and development investment in new medicines by the biopharmaceutical industry stood at $65.2 billion in 2008. A few examples of biologics made with recombinant DNA technology include:

USAN/INN	Trade name	Indication	Technology	Mechanism of action
abatacept	Orencia	rheumatoid arthritis	immunoglobin CTLA-4 fusion protein	T-cell deactivation
adalimumab	Humira	rheumatoid arthritis, ankylosing spondylitis, psoriatic arthritis, psoriasis, Ulcerative Colitis, Crohn's disease	monoclonal antibody	TNF antagonist
alefacept	Amevive	chronic plaque psoriasis	immunoglobin G1 fusion protein	incompletely characterized
erythropoietin	Epogen	anemia arising from cancer chemotherapy, chronic renal failure, etc.	recombinant protein	stimulation of red blood cell production
etanercept	Enbrel	rheumatoid arthritis, ankylosing spondylitis, psoriatic arthritis, psoriasis	recombinant human TNF-receptor fusion protein	TNF antagonist
infliximab	Remicade	rheumatoid arthritis, ankylosing spondylitis, psoriatic arthritis, psoriasis, Ulcerative Colitus, Crohn's disease	monoclonal antibody	TNF antagonist

trastuzumab	Herceptin	breast cancer	humanized monoclonal antibody	HER2/neu (erbB2) antagonist
ustekinumab	Stelara	psoriasis	humanized monoclonal antibody	IL-12 and IL-23 antagonist
denileukin diftitox	Ontak	cutaneous T-cell lymphoma (CTCL)	Diphtheria toxin engineered protein combining Interleukin-2 and Diphtheria toxin	Interleukin-2 receptor binder
golimumab	Simponi	rheumatoid arthritis, psoriatic arthritis, ankylosing spondylitis, Ulcerative colitis	monoclonal antibody	TNF antagonist

Vaccines

Many vaccines are grown in tissue cultures.

Gene Therapy

Viral gene therapy involves artificially manipulating a virus to include a desirable piece of genetic material.

Biosimilars

With the expiration of numerous patents for blockbuster biologics between 2012 and 2019, the interest in biosimilar production, i.e., follow-on biologics, has increased. Compared to small molecules that consist of chemically identical active ingredients, biologics are vastly more complex and consist of a multitude of subspecies. Due to their heterogeneity and the high process sensitivity, neither originators nor follow-on manufacturers produce reliably constant quality profiles over time. The process variations are monitored by modern analytical tools (e.g., liquid chromatography, immunoassays, mass spectrometry, etc.) and describe a unique design space for each biologic.

Thus, biosimilars require a different regulatory framework compared to small-molecule generics. Legislation in the 21st century has addressed this by recognizing an intermediate ground of testing for biosimilars. The filing pathway requires more testing than for small-molecule generics, but less testing than for registering completely new therapeutics.

In 2003, the European Medicines Agency introduced an adapted pathway for biosimilars, termed *similar biological medicinal products*. This pathway is based on a thorough demonstration of "comparability" of the "similar" product to an existing approved product. Within the United States, the Patient Protection and Affordable Care Act of 2010 created an abbreviated approval pathway for biological products shown to be biosimilar to, or interchangeable with, an FDA-licensed reference biological product. A major hope linked to the introduction of biosimilars is a reduction of costs to the patients and the healthcare system.

Commercialization

When a new biopharmaceutical is developed, the company will typically apply for a patent, which is a grant for exclusive manufacturing rights. This is the primary means by which the developer of the drug can recover the investment cost for development of the biopharmaceutical. The patent laws in the United States and Europe differ somewhat on the requirements for a patent, with the European requirements are perceived as more difficult to satisfy. The total number of patents granted for biopharmaceuticals has risen significantly since the 1970s. In 1978 the total patents granted was 30. This had climbed to 15,600 in 1995, and by 2001 there were 34,527 patent applications.

Large-scale Production

Biopharmaceuticals may be produced from microbial cells (e.g., recombinant *E. coli* or yeast cultures), mammalian cell lines and plant cell cultures and moss plants in bioreactors of various configurations, including photo-bioreactors. Important issues of concern are cost of production (low-volume, high-purity products are desirable) and microbial contamination (by bacteria, viruses, mycoplasma). Alternative platforms of production which are being tested include whole plants (plant-made pharmaceuticals).

Transgenics

A potentially controversial method of producing biopharmaceuticals involves transgenic organisms, particularly plants and animals that have been genetically modified to produce drugs. This production is a significant risk for the investor, due to production failure or scrituny from regulatory bodies based on perceived risks and ethical issues. Biopharmaceutical crops also represent a risk of cross-contamination with non-engineered crops, or crops engineered for non-medical purposes.

One potential approach to this technology is the creation of a transgenic mammal that can produce the biopharmaceutical in its milk, blood, or urine. Once an animal is produced, typically using the pronuclear microinjection method, it becomes efficacious to use cloning technology to create additional offspring that carry the favorable modified genome. The first such drug manufactured from the milk of a genetically modified goat was ATryn, but marketing permission was blocked by the European Medicines Agency in February 2006. This decision was reversed in June 2006 and approval was given August 2006.

Regulation

European Union

In the European Union, a biological medicinal product is one of the active substance(s) produced from or extracted from a biological (living) system, and requires, in addition to physico-chemical testing, biological testing for full characterisation. The characterisation of a biological medicinal product is a combination of testing the active substance and the final medicinal product together with the production process and its control. For example:

- Production process – it can be derived from biotechnology or from other technologies. It may be prepared using more conventional techniques as is the case for blood or plasma-derived products and a number of vaccines.

- Active substance – consisting of entire microorganisms, mammalian cells, nucleic acids, proteinaceous, or polysaccharide components originating from a microbial, animal, human, or plant source.

- Mode of action – therapeutic and immunological medicinal products, gene transfer materials, or cell therapy materials.

United States

In the United States, biologics are regulated by the FDA's Center for Biologics Evaluation and Research (CBER) whereas drugs are regulated by the Center for Drug Evaluation and Research. Approval may require several years of clinical trials, including trials with human volunteers. Even after the drug is released, it will still be monitored for performance and safety risks. The manufacture process must satisfy the FDA's "Good Manufacturing Practices", which are typically manufactured in a clean room environment with strict limits on the amount of airborne particles.

References

- Peter J. Russel (2005). iGenetics: A Molecular Approach. San Francisco, California, United States of America: Pearson Education. ISBN 0-8053-4665-1.

- Rantala, Milgram, M., Arthur (1999). Cloning: For and Against. Chicago, Illinois: Carus Publishing Company. p. 1. ISBN 0-8126-9375-2.

- de Grey, Aubrey; Rae, Michael (September 2007). Ending Aging: The Rejuvenation Breakthroughs that Could Reverse Human Aging in Our Lifetime. New York, NY: St. Martin's Press, 416 pp. ISBN 0-312-36706-6.

- McHugen, Alan (14 September 2000). "Chapter 1: Hors-d'oeuvres and entrees/What is genetic modification? What are GMOs?". Pandora's Picnic Basket. Oxford University Press. ISBN 978-0198506744.

- Head, Graham; Hull, Roger H; Tzotzos, George T. (2009). Genetically Modified Plants: Assessing Safety and Managing Risk. London: Academic Pr. p. 244. ISBN 0-12-374106-8.

- Tan, WS.; Carlson, DF.; Walton, MW.; Fahrenkrug, SC.; Hackett, PB. (2012). "Precision editing of large animal genomes". Adv Genet. Advances in Genetics. 80: 37–97. doi:10.1016/B978-0-12-404742-6.00002-8. ISBN 9780124047426. PMC 3683964. PMID 23084873.

- Avise, John C. (2004). The hope, hype & reality of genetic engineering: remarkable stories from agriculture, industry, medicine, and the environment. Oxford University Press US. p. 22. ISBN 978-0-19-516950-8.

- Reece, Jane B.; Urry, Lisa A.; Cain, Michael L.; Wasserman, Steven A.; Minorsky, Peter V.; Jackson, Robert B. (2011). Campbell Biology Ninth Edition. San Francisco: Pearson Benjamin Cummings. p. 421. ISBN 0-321-55823-5.

- Watson, James D. (2007). Recombinant DNA: Genes and Genomes: A Short Course. San Francisco: W.H. Freeman. ISBN 0-7167-2866-4.

- Russell, David W.; Sambrook, Joseph (2001). Molecular cloning: a laboratory manual. Cold Spring Harbor, N.Y: Cold Spring Harbor Laboratory. ISBN 0-87969-576-5.

- Nerem, R.M. (2000). Vacanti, Joseph; Lanza, R. P.; Langer, Robert S., eds. Principles of tissue engineering (2nd ed.). Boston: Academic Press. ISBN 0-12-436630-9.

- Zastrow, Mark (8 February 2016). "Inside the cloning factory that creates 500 new animals a day". New Scientist. Retrieved 23 February 2016.

- Shukman, David (14 January 2014) China cloning on an 'industrial scale' BBC News Science and Environment, Retrieved 27 February 2016

- Baer, Drake (8 September 2015). "This Korean lab has nearly perfected dog cloning, and that's just the start". Tech Insider, Innovation. Retrieved 27 February 2016.

- Pollack, Andrew (19 November 2015). "Genetically Engineered Salmon Approved for Consumption". The New York Times. Retrieved 21 April 2016.

- Angulo, E.; Cooke, B. (2002). "First synthesize new viruses then regulate their release? The case of the wild rabbit". Molecular Ecology. 11: 2703–9. doi:10.1046/j.1365-294X.2002.01635.x. PMID 12453252. Retrieved 16 May 2016.

- Adams; et al. (2 August 2002). "The Case for Genetic Engineering of Native and Landscape Trees against Introduced Pests and Diseases". Conservation Biology. 16: 874–879. doi:10.1046/j.1523-1739.2002.00523.x. Retrieved 16 May 2016.

- Thomas; et al. (25 September 2013). "Ecology: Gene tweaking for conservation". Nature. 501: 485–6. doi:10.1038/501485a. PMID 24073449. Retrieved 16 May 2016.

- Thomas, Kate; Pollack, Andrew (15 July 2015). "Specialty Pharmacies Proliferate, Along With Questions". Sinking Spring, Pa.: New York Times. Retrieved 5 October 2015.

5

Biotechnological Products

The products developed from the use of living systems and organisms are known as biotechnological products. Some of the biological products discussed in the chapter are bioSteel, biopolymer, burton, arctic apples and specialty drugs. The content provides an overview of the subject matter incorporating all the major products of biotechnology.

BioSteel

BIOSTEEL is an artificial silk fibre made from recombinantly produced silk proteins. The silk technology has been developed by AMSilk GmbH, a German industrial biotechnology company. The raw material for fibre production is manufactured via a proprietary industrial biotechnology production process.

BIOSTEEL has many special properties, such as high toughness and elasticity and a high E-modulus. It thus enables light weight construction. Moreover the fiber is 100% biodegradable and bacteriostatic.

BioSteel was a trademark name for a high-strength based fiber material made of the recombinant spider silk-like protein extracted from the milk of transgenic goats, made by Nexia Biotechnologies, and later by the Randy Lewis lab of the University of Wyoming and Utah State University. It is reportedly 7-10 times as strong as steel if compared for the same weight, and can stretch up to 20 times its unaltered size without losing its strength properties. It also has very high resistance to extreme temperatures, not losing any of its properties within -20 to 330 degrees Celsius.

The company had created lines of goats to produce recombinant versions of either the MaSpI (Major ampullate spidroin I) or dragline I (for its superious elasticity, flexibility and strength) from *Nephila clavipes*, the golden orb weaver) or MaSpII (Major ampullate spidroin 2 or dragline 2 from *Nephila clavipes*) dragline proteins in their milk. When the female goats lactate, the milk, containing the recombinant DNA silk, was to be harvested and subjected to chromatographic techniques to purify the recombinant silk proteins.

The purified silk proteins could be dried, dissolved using solvents (DOPE formation) and transformed into microfibers using wet-spinning fiber production methods. The spun fibers were reported to have tenacities in the range of 2 - 3 grams/denier and elongation range of 25-45%. The "Biosteel biopolymer" had been transformed into nanofibers and nanomeshes using the electrospinning technique.

Nexia is the only company which has successfully produced fibres from spider silk expressed in goat's milk. The Lewis lab has produced fibers from recombinant spider silk protein and synthetic spider silk proteins and chimeras produced in both recombinant E. coli and the milk of recombinant goats, however, no one has been able to produce the silk in commercial quantities thus far. The Company was founded in 1993 by Dr. Jeffrey Turner and Paul Ballard, and was sold in 2005 to Pharmathene.

In 2009, two transgenic goats were sold to the Canada Agriculture Museum after Nexia Biotechnologies went bankrupt.

Research has since continued with the help of Randy Lewis, a professor formerly at the University of Wyoming and now at Utah State University. He was also able to successfully breed spider goats in order to create artificial silk. There are now about 30 spider goats at a university-run farm.

Applications of artificial spider silk biopolymers include using it for the coating of all kinds of implants and medical products as well as for artificial ligaments and tendons due to its elastic tendencies and also since it is a natural product which will synthesize well with the body. Furthermore, artificial silk biopolymers can be applicated in personal care products as well as in textile products..

Biopolymer

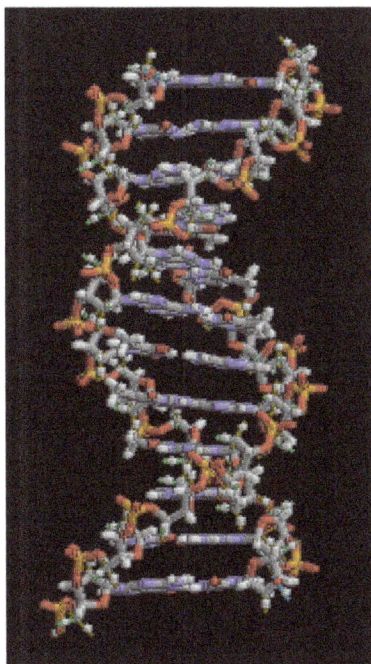

In the structure of DNA is a pair of **biopolymers**, polynucleotides, forming the double helix

Biopolymers are polymers produced by living organisms; in other words, they are polymeric biomolecules. Since they are polymers, biopolymers contain monomeric units that are covalently bonded to form larger structures. There are three main classes of biopolymers, classified according to the monomeric units used and the structure of the biopolymer formed: polynucleotides (RNA and DNA), which are long polymers composed of 13 or more nucleotide monomers; polypeptides, which are short polymers of amino acids; and polysaccharides, which are often linear bonded polymeric carbohydrate structures.

IUPAC Definition

Substance composed of one type of *biomacromolecules*.

Note 1: Modified from the definition given in ref. in order to avoid confusion between *polymer* and *macromolecule* in the fields of proteins, polysaccharides, polynucleotides, and bacterial aliphatic polyesters.

Note 2: The use of the term "biomacromolecule" is recommended when molecular characteristics are considered.

Cellulose is the most common organic compound and biopolymer on Earth. About 33 percent of all plant matter is cellulose. The cellulose content of cotton is 90 percent, for wood is 50 percent.

Biopolymers vs Synthetic Polymers

A major defining difference between biopolymers and other polymers can be found in their structures. All polymers are made of repetitive units called monomers. Biopolymers often have a well-defined structure, though this is not a defining characteristic (example: lignocellulose): The exact chemical composition and the sequence in which these units are arranged is called the primary structure, in the case of proteins. Many biopolymers spontaneously fold into characteristic compact shapes, which determine their biological functions and depend in a complicated way on their primary structures. Structural biology is the study of the structural properties of the biopolymers. In contrast, most synthetic polymers have much simpler and more random (or stochastic) structures. This fact leads to a molecular mass distribution that is missing in biopolymers. In fact, as their synthesis is controlled by a template-directed process in most *in vivo* systems, all biopolymers of a type (say one specific protein) are all alike: they all contain the similar sequences and numbers of monomers and thus all have the same mass. This phenomenon is called monodispersity in contrast to the polydispersity encountered in synthetic polymers. As a result, biopolymers have a polydispersity index of 1.

Conventions and Nomenclature

Polypeptides

The convention for a polypeptide is to list its constituent amino acid residues as they occur from the amino terminus to the carboxylic acid terminus. The amino acid residues are always joined by peptide bonds. Protein, though used colloquially to refer to any polypeptide, refers to larger or fully functional forms and can consist of several polypeptide chains as well as single chains. Proteins can also be modified to include non-peptide components, such as saccharide chains and lipids.

Nucleic Acids

The convention for a nucleic acid sequence is to list the nucleotides as they occur from the 5' end to the 3' end of the polymer chain, where 5' and 3' refer to the numbering of carbons around the ribose ring which participate in forming the phosphate diester linkages of the chain. Such a sequence is called the primary structure of the biopolymer.

Sugars

Sugar-based biopolymers are often difficult with regards to convention. Sugar polymers can be linear or branched and are typically joined with glycosidic bonds. The exact placement of the linkage can vary, and the orientation of the linking functional groups is also important, resulting in α- and β-glycosidic bonds with numbering definitive of the linking carbons' location in the ring. In addition, many saccharide units can undergo various chemical modifications, such as amination, and can even form parts of other molecules, such as glycoproteins.

Structural Characterization

There are a number of biophysical techniques for determining sequence information. Protein sequence can be determined by Edman degradation, in which the N-terminal residues are hydrolyzed from the chain one at a time, derivatized, and then identified. Mass spectrometer techniques can also be used. Nucleic acid sequence can be determined using gel electrophoresis and capillary electrophoresis. Lastly, mechanical properties of these biopolymers can often be measured using optical tweezers or atomic force microscopy. Dual polarization interferometry can be used to measure the conformational changes or self-assembly of these materials when stimulated by pH, temperature, ionic strength or other binding partners.

Biopolymers as Materials

Some biopolymers- such as (PLA), naturally occurring zein, and poly-3-hydroxybutyrate can be used as plastics, replacing the need for polystyrene or polyethylene based plastics.

Some plastics are now referred to as being 'degradable', 'oxy-degradable' or 'UV-degradable'. This means that they break down when exposed to light or air, but these plastics are still primarily (as much as 98 per cent) oil-based and are not currently certified as 'biodegradable' under the European Union directive on Packaging and Packaging Waste (94/62/EC). Biopolymers will break down, and some are suitable for domestic composting.

Biopolymers (also called renewable polymers) are produced from biomass for use in the packaging industry. Biomass comes from crops such as sugar beet, potatoes or wheat: when used to produce biopolymers, these are classified as non food crops. These can be converted in the following pathways:

Sugar beet > Glyconic acid > Polyglyconic acid

Starch > (fermentation) > Lactic acid > Polylactic acid (PLA)

Biomass > (fermentation) > Bioethanol > Ethene > Polyethylene

Many types of packaging can be made from biopolymers: food trays, blown starch pellets for shipping fragile goods, thin films for wrapping.

Environmental Impacts

Biopolymers can be sustainable, carbon neutral and are always renewable, because they are made from plant materials which can be grown indefinitely. These plant materials come from agricultural non food crops. Therefore, the use of biopolymers would create a sustainable industry. In contrast, the feedstocks for polymers derived from petrochemicals will eventually deplete. In addition, biopolymers have the potential to cut carbon emissions and reduce CO_2 quantities in the atmosphere: this is because the CO_2 released when they degrade can be reabsorbed by crops grown to replace them: this makes them close to carbon neutral.

Biopolymers are biodegradable, and some are also compostable. Some biopolymers are biodegradable: they are broken down into CO_2 and water by microorganisms. Some of these biodegradable biopolymers are compostable: they can be put into an industrial composting process and will break down by 90% within six months. Biopolymers that do this can be marked with a 'compostable' symbol, under European Standard EN 13432 (2000). Packaging marked with this symbol can be put into industrial composting processes and will break down within six months or less. An example of a compostable polymer is PLA film under 20µm thick: films which are thicker than that do not qualify as compostable, even though they are biodegradable. In Europe there is a home composting standard and associated logo that enables consumers to identify and dispose of packaging in their compost heap.

Burton

People

- Burton (name)

Places

Australia

- Burton, South Australia, a suburb of Adelaide

Canada

- Burton, British Columbia
- Burton, New Brunswick
- Burton, Ontario
- Burtons, Nova Scotia
- Burton, Durham Region, Ontario
- Burton, Parry Sound District, Ontario
- Burton, Prince Edward Island
- Lac-Burton, Quebec

- Burton Lake, Saskatchewan

United Kingdom

England

- Burton, Chester, near Tarvin in Cheshire
- Burton, Gowy, near Tarporley in Cheshire
- Burton, Neston, Cheshire, on the Wirral Peninsula in Cheshire
- Burton in Lonsdale, near Ingleton in North Yorkshire
- Burton-in-Kendal, Cumbria
- Burton, Dorset
- Burton, Lincolnshire
- Burton-upon-Stather, North Lincolnshire
- Burton, Northumberland
- Burton upon Trent, Staffordshire
    - o Burton (UK Parliament constituency), covering Burton upon Trent and surrounding areas
- Burton, Sussex
- Burton, Wiltshire, near Chippenham
- Burton, Mere, Wiltshire

Wales

- Burton, Wrexham
- Burton, Pembrokeshire

United States

- Burton, Arizona
- Burton, Arkansas
- Burton, Georgia
- Burton, Idaho
- Burton, Illinois
- Burton, Kentucky
- Burton, Louisiana
- Burton, Michigan (Genesee County)

- Burton, Shiawassee County, Michigan
- Burton, Mississippi
- Burton, Howard County, Missouri
- Burton, Wayne County, Missouri
- Burton, Montana
- Burton, Nebraska
- Burton, North Carolina
- Burton, Ohio
- Burton, Cimarron County, Oklahoma
- Burton, Hughes County, Oklahoma
- Burton, Noble County, Oklahoma
- Burton, South Carolina
- Burton, Tennessee
- Burton, Johnson County, Texas
- Burton, Washington County, Texas
- Burton, Utah
- Burton, Clark County, Washington
- Burton, King County, Washington
- Burton, West Virginia
- Burton, Wisconsin

Business

- Burton Brothers, a New Zealand photographic firm
- Burton Snowboards, a snowboard, snow sportswear, and streetwear manufacturer
- Burton (retailer), a menswear retailer in the UK
- Burton (car), a Dutch sports car
- The Burton Store, in Tappan, New York, USA
- Burton's Biscuit Company, a biscuit producer in the UK

Other

- Burton process, a thermal cracking process invented by William M. Burton, used to produce diesel fuel

- Burton (nut), edible nut in the Hican family

- Burton (car), a Dutch sports car

- HMS *Burton* (L08), the original name, changed prior to her completion, of the British escort destroyer HMS *Exmoor* (L08) which served in the Royal Navy from 1941 to 1945

Arctic Apples

Arctic Apples are a suite of trademarked apples that contain a nonbrowning trait (when the apples are subjected to mechanical damage, such as slicing or bruising, the apple flesh remains its original color) introduced through biotechnology. They were developed through a process of genetic engineering and precision breeding by Okanagan Specialty Fruits Inc. (OSF). Specifically, gene silencing reduces the expression of polyphenol oxidase (PPO), thus preventing the fruit from browning. It is the first approved food product to use that technique. This is unlike many other genetically modified foods, which insert genes from other species for the purposes of pesticide tolerance or insect resistance.

Okanagan Specialty Fruits has petitioned for regulatory approval for two apple varieties in Canada from the Canadian Food Inspection Agency (CFIA) and Health Canada and in the US from the Animal and Plant Health Inspection Service (APHIS), part of the United States Department of Agriculture (USDA). In 2012, a field test application was approved to conduct a 20-acre study of the apple in the state of Washington. The apples were approved by the USDA in February 2015 and by the FDA in March 2015, becoming the first genetically modified apple approved for US sale.

Specialty Drugs (United States)

Specialty drugs or specialty pharmaceuticals are a recent designation of pharmaceuticals that are classified as high-cost, high complexity and/or high touch. Specialty drugs are often biologics—"drugs derived from living cells" that are injectable or infused (although some are oral medications). They are used to treat complex or rare chronic conditions such as cancer, rheumatoid arthritis, hemophilia, H.I.V. psoriasis, inflammatory bowel disease (IBD) and Hepatitis C. In 1990 there were 10 specialty drugs on the market, in the mid-1990s there were fewer than 30, by 2008 there were 200, and by 2015 there were 300. Drugs are often defined as specialty because their price is much higher than that of non-specialty drugs. Medicare defines any drug for which the negotiated price is $600 per month or more, as a specialty drug which is placed in a specialty tier that requires a higher patient cost sharing. Drugs are also identified as specialty when there is a special handling requirement or the drug is only available via a limited distributions network. By 2015 "specialty medications accounted for one-third of all spending on drugs in the United States, up from 19 percent in 2004 and heading toward 50 percent in the next 10 years," according to IMS Health, which tracks prescriptions." According to a 2010 article in Forbes, specialty drugs for rare diseases became more expensive "than anyone imagined" and their success came "at a time when the traditional drug business of selling medicines to the masses" was "in decline." In 2015 analysis by *The Wall Street Journal* suggested the large premium was due to the perceived value of rare disease treatments which usually are very expensive when compared to treatments for more common diseases.

Definition and Common Characteristics

Medications must be either identified as high cost, high complexity or high touch to be classified as a specialty medication by Magellan Rx Management. Specialty pharmaceuticals are defined as "high-cost oral or injectable medications used to treat complex chronic conditions." According to a 2013 article in the *Journal of Managed Care & Specialty Pharmacy*, on the increasingly important role of specialty drugs in the treatment of chronic conditions and their cost, drugs are most typically defined as specialty because they are expensive. Other criteria used to define a drug as specialty include "biologic drugs, the need to inject or infuse the drug, the requirement for special handling, or drug availability only via a limited distribution network." The price of specialty drugs compared to non-specialty drugs is very high, "more than $1,000 per 30-day supply."

Specialty drugs cover over forty therapeutic categories and special disease states with over 500 drugs.

Vogenberg claims that there is no standard definition of a specialty drug which is one of the reasons they are difficult to manage. "[T]hose pharmaceuticals that usually require special handling, administration, unique inventory management, and a high level of patient monitoring and support to consumers with specific chronic conditions, acute events, or complex therapies, and provides comprehensive patient education services and coordination with the patient and prescribe."

High Cost

Drugs are most typically defined as specialty because they are expensive. They are high cost "both in total and on a per-patient basis." High-cost medications are typically priced at more than $1,000 per 30-day supply. The Medicare Part D program "defines a specialty drug as one that costs more than $600 per month." Most of the prescriptions filled by Pennsylvania-licensed Philidor Rx Services Pennsylvania-licensed specialty online mail-order pharmacy, which mainly sold Valeant Pharmaceuticals International Inc expensive drugs directly to patients and handled insurance claims on the customers' behalf, such as Solodyn, Jublia, and Tretinoin

High Complexity

Specialty drugs are more complex to manufacture. They are "highly complex medications, typically biology-based, that structurally mimic compounds found within the body." Specialty drugs are often biologics—"drugs derived from living cells"—but biologics are "not always deemed to be specialty drugs." Biologics "may be produced by biotechnology methods and other cutting-edge technologies. Gene-based and cellular biologics, for example, often are at the forefront of biomedical research, and may be used to treat a variety of medical conditions for which no other treatments are available."

"In contrast to most drugs that are chemically synthesized and their structure is known, most biologics are complex mixtures that are not easily identified or characterized. Biological products, including those manufactured by biotechnology, tend to be heat sensitive and susceptible to microbial contamination. Therefore, it is necessary to use aseptic principles from initial manufacturing steps, which is also in contrast to most conventional drugs. Biological products often represent the cutting-edge of biomedical research and, in time, may offer the most effective means to treat a variety of medical illnesses and conditions that presently have no other treatments available."

— U.S. Food and Drug Administration

According to the U.S. Food and Drug Administration (FDA) biologics, or

"Biological products include a wide range of products such as vaccines, blood and blood components, allergenics, somatic cells, gene therapy, tissues, and recombinant therapeutic proteins. Biologics can be composed of sugars, proteins, or nucleic acids or complex combinations of these substances, or may be living entities such as cells and tissues. Biologics are isolated from a variety of natural sources - human, animal, or microorganism..."

— U.S. Food and Drug Administration

High Touch

Some specialty drugs can be oral medications or self-administered injectables. Others may be professionally administered or injectables/infusions. High-touch patient care management is usually required to control side effects and ensure compliance. Specialized handling and distribution are also necessary to ensure appropriate medication administration. Specialty drugs patient care management is meant to be both high technology and high touch care, or Patient-centered care (PCC) with "more face-to-face time, more personal connections." PCC is defined by the Institute of Medicine as "care that is respectful of and responsive to individual patient preferences, needs and values."

Specialty drugs may be "difficult for patients to take without ongoing clinical support."

Limited Availability

Specialty drugs might have special requirements for handling procedures and administration including the necessity of having controlled environments such as highly specific temperature controls to ensure product integrity. They are often only available via a limited distributions network such as a special pharmacy. Specialty drugs may be "challenging for providers to manage."

Rare and Complex Diseases

Specialty drugs may be taken "by relatively small patient populations presenting with complex medical conditions."

History

"Specialty pharmacies have their roots in the 1970s, when they began delivering temperature-controlled drugs to treat cancer, HIV, infertility and hemophilia."

"The business grew as more drugs became available for patients to inject themselves and as insurers sought to manage expenses for patients with chronic conditions, according to a report from IMS Health. Manufacturers have increasingly relied on these pharmacies when it comes to fragile medicines that need special handling or have potentially dangerous side effects that require them to be taken under a management program."

—Bloomberg 2015

According to the *The American Journal of Managed Care,* in 1990 there were 10 specialty drugs on the market; According to the National Center for Biotechnology Information, by the mid 1990s, there were fewer than 30 specialty drugs on the market, but by 2008 that number had increased to 200.

Specialty drugs may also be designated as orphan drugs or ultra-orphan drugs under the U. S. Orphan Drug Act of 1983 (ODA). The ODA was enacted to facilitate development of orphan drugs—drugs for rare diseases such as Huntington's Disease, myoclonus, ALS, Tourette syndrome and muscular dystrophy which affect small numbers of individuals residing in the United States.

Not all specialty drugs are orphan drugs. According to Thomson Reuters in their 2012 publication "The Economic Power of Orphan Drugs", there has been increased investing in orphan drug Research and Development partly since the U.S. Congress enacted the ODA giving an extra monopoly for drugs for "orphan diseases" that hit fewer than 200,000 people in the country. Similar Acts came into existence in other regions of the world many driven by "high-profile philanthropic funding." According to a 2010 article in Forbes prior to 1983 drug companies largely ignored rare diseases and focused on drugs that affected millions of patients.

The term specialty drugs was used as early as 1988 in a New York Times article about Eastman Kodak Company's acquisition of the New York-based Sterling Drug Inc., maker of specialty drugs along with many and diverse other products. In 2000 when Shire Pharmaceuticals acquired Bio-Chem Pharma in 2000 they created a specialty pharmaceuticals company. By 2001 Shire was one the fastest growing specialty pharmaceutical companies in the world.

By 2001 CVS' specialty pharmacy ProCare was the "largest integrated retail/mail provider of specialty pharmacy services" in the United States. It was consolidated with their pharmacy benefit management company, PharmaCare in 2002 to In their 2001 annual report CVS anticipated that the "$16 billion specialty pharmacy market" would grow at "an even faster rate than traditional pharmacy due in large part to the robust pipeline of biotechnology drugs." By 2014 CVS Caremark, Express Scripts and Walgreens represented more than 50% of the specialty drug market in the United States.

When an increasing number of oral oncology agents first entered the market between 2000 and 2010, most cancer care was provided in a community oncology practices. By 2008 many other drugs had been developed to treat cancer, and drug development had exploded into a multibillion-dollar industry.

In 2003 the Medicare Prescription Drug, Improvement, and Modernization Act was enacted - the largest overhaul of Medicare in the public health program's 38-year history—included Medicare Part D an entitlement benefit for prescription drugs, through tax breaks and subsidies. In 2004 the U. S. Centers for Medicare and Medicaid Services (CMS) prepared a report on final guidance regarding access to drug coverage enacted under in which they included the specialty drugs tier in the prescription drug formulary. At that time CMS guidelines included four tiers: tier 1 includes preferred generics, tier 2 includes preferred brands, tier 3 includes non-preferred brands and generics and tier 4 included specialty drugs. By January 1, 2006 the controversial Medicare Part D was put in effect. It was a massive expansion of the federal government's provision of prescription drug coverage to previously uninsured Americans, particularly seniors. In 2006 in the United States there was no standard nomenclature, so sellers could call the plan anything they wanted and cover whatever drugs they wanted.

By 2008 most prescription medication plans in the United States used specialty drug tiers, and some had a separate benefit tier for injectable drugs. Beneficiary cost sharing was higher for drugs in these tiers.

By 2011 in the United States a growing number of Medicare Part D health insurance plans—which normally includes generic, preferred, and non-preferred tiers with an accompanying rate of cost-sharing or co-payment—had added an "additional tier for high-cost drugs which is referred to as a specialty tier."

By 2014 in the United States, in the new Health Insurance Marketplace—following the implementation of the U.S. Health Care Reform Act, also known as Obamacare—most health plans had a four- or five-tier prescription drug formulary with specialty drugs in the highest of the tiers.

AARP

According to an AARP 2015 report, "All but 4 of the 46 therapeutic categories of specialty drug products had average annual retail price increases that exceeded the rate of general inflation in 2013. Price increases by therapeutic category ranged from 1.7 percent to 77.2 percent."

Risk Evaluation and Mitigation Strategies (REMS)

On September 27, 2007 President George W. Bush amended the Food and Drug Administration Amendments Act of 2007 (FDAAA) to authorize the FDA to require "risk evaluation and mitigation strategies (REMS) on medications if necessary to minimize the risks associated with some drugs." These medications were designated as specialty drugs and required specialty pharmacies. When the FDA approves a new drug they may require a REMS program which "may contain any combination of 5 criteria: Medication Guide, Communication Plan, Elements to Assure Safe Use, Implementation System, and Timetable for Submission of Assessments." "In 2010, 48% of all new molecular entities, and 60% of all new specialty drug approvals, required a REMS program." Risk-reduction mechanisms can include the "use of specialized distribution partners," special pharmacy.

Breakthrough Therapy

In 2013 the FDA introduced the breakthrough therapy designation program which cut the development process of new therapies by several years. This meant that the FDA could "introduce important medicines to the market based on very promising phase 2 rather than phase 3 clinical trial results." Shortly after the law was enacted, Ivacaftor, in January 2013, became the first drug to receive the breakthrough therapy designation.

On February 3, 2015 New York-based Pfizer's drug Ibrance was approved through the FDA's Breakthrough Therapy designation program as a treatment for advanced breast cancer. It can only be ordered through "through select" specialty pharmacies and "sells for $9,850 for 30 days or $118,200 for a year's supply before discounts." According to a statement by the New York-based Pfizer the price "is not the cost that most patients or payors pay" since most prescriptions are dispensed through health plans, which negotiate discounts for medicines or get government-mandated price concessions.

Trends in Spending in the United States

According to Express Scripts,

"[T]he pharmacy landscape [in the United States] underwent a seismic change, and the budgetary impact to healthcare payers was significant. U.S. prescription drug spend increased 13.1% in 2014 – the largest annual increase since 2003 – and this was largely driven by an unprecedented 30.9% increase in spending on specialty medications. Utilization of traditional medications stayed flat (-0.1%), while the use of specialty drugs increased 5.8%. The largest factors contributing to the increased spending, however, were the price increases for these medication categories – 6.5% for traditional and 25.2% for specialty. While specialty medications represent only 1% of all U.S. prescriptions, these medications represented 31.8% of all 2014 drug spend – an increase from 27.7% in 2013."

—Express Scripts Drug Trend Report

By 2015 "specialty medications account for one-third of all spending on drugs in the United States, up from 19 percent in 2004 and heading toward 50 percent in the next 10 years, according to IMS Health, which tracks prescriptions." The specialty pharmacy business had $20 billion in sales in 2005. By 2014 it had grown to "$78 billion in sales." In Canada by 2013 "specialty drugs made up less than 1.3 percent of all Canadian prescriptions, but accounted for 24 percent of Canada's total spending on prescription drugs."

When Randy Vogenberg of the Institute for Integrated Healthcare in Massachusetts and a co-leader of the Midwest Business Group initiative, began investigating specialty drugs in 2003, it "wasn't showing up on the radar." By 2009 specialty drugs had started doubling in cost and payers such as employers began to question. Vogenberg observed that by 2014 health care reform had changed the landscape for specialty drugs. There is a shift away from a marketplace based on a predominately clinical perspective, to one that puts economics first and clinical second.

Insurance Payer Definition

In the United States, private insurance payers will favour a lower-cost agent preferring generics and biosimilars to the more expensive specialty drugs if there is no peer-reviewed or evidence-based justification for them.

According to a 2012 report by Sun Life Financial the average cost of specialty drug claims was $10,753 versus $185 for non-specialty drugs and the cost of specialty drugs continues to rise. With such steep prices by 2012 specialty drugs represented 15-20% of prescription drug reimbursement claims.

Patient advocacy groups that lobby for payment for specialty drugs include the Alliance for Patient Access (AfPA), formed in 2006 and which according to a 2014 article in the Wall Street Journal "represents physicians and is largely funded by the pharmaceutical industry. The contributors mostly include brand-name drug makers and biotechs, but some – such as Pfizer and Amgen – are also developing biosimilars."

In 2013 AfPA director David Charles published an article on specialty drugs in which he agreed with the findings of the Congressional Budget Office that spending on prescription medications "saves costs in other areas of healthcare spending." He observed that specialty drugs are so high priced that many patients do not fill prescriptions resulting in more serious health problems increasing. His article referred to specialty drugs such as "new cancer drugs specially formulated for patients with specific genetic markers." He explained the high cost of these "individualized medications based on diagnostic testing; and "biologics," or medicines created through biologic processes, rather than chemically synthesized like most pharmaceuticals." He argued that there should be a slight increase in co-pays for the more commonly using lower-tier medications to allow a lower co-pay for those who "require high-cost specialty tier medications."

Top Specialty Therapy Classes and Average Prescription Costs

According to the 2014 Express Scripts Drug Trend Report, the most significant increase in prescription drugs in the United States in 2014 was due to "increased inflation and utilization of hepatitis C and compounded medications." "Excluding those two therapy classes, overall drug spend would have increased only 6.4%.

The cost of "the top three specialty therapy classes – inflammatory conditions, multiple sclerosis and oncology – contributed 55.9% of the spend for all specialty medications billed through the pharmacy benefit in 2014. The U.S. spent 742.6% more on hepatitis C medications in 2014 than it did in 2013; this therapy class was not among the top 10 specialty classes in 2013.

Specialty Pharmacies

As the market demanded specialization in drug distribution and clinical management of complex therapies, specialized pharma (SP) evolved. By 2001 CVS' specialty pharmacy ProCare was the "largest integrated retail/mail provider of specialty pharmacy services" in the United States. It was consolidated with their pharmacy benefit management company, PharmaCare in 2002 to In their 2001 annual report CVS anticipated that the "$16 billion specialty pharmacy market" would grow at "an even faster rate than traditional pharmacy due in large part to the robust pipeline of biotechnology drugs." By 2014 CVS Caremark, Express Scripts and Walgreens represented more than 50% of the specialty drug market in the United States.

The specialty pharmacy business had $20 billion in sales in 2005. By 2014 it had grown to "$78 billion in sales."

Specialty pharmacies came into existence to as a result of unmet needs. According to the National Comprehensive Cancer Network the "primary goals of specialty pharmacies are to ensure the appropriate use of medications, maximize drug adherence, enhance patient satisfaction through direct interaction with healthcare professionals, minimize cost impact, and optimize pharmaceutical care outcomes and delivery of information."

McKesson Specialty Care Solutions, a division of McKesson Corporation, is "one of the largest distributors of specialty drugs, biologics and rheumatology drugs to community-based specialty practices." It is "a leader in the development, implementation and management of FDA-mandated Risk Evaluation and Mitigation Strategies (REMS) for manufacturers." For example in order ProStrakan Group plc, an international pharmaceutical company based in the UK works with McKesson Specialty Care Solutions to administer its FDA-approved Risk Evaluation and Mitigation Strategy (REMS) program for Abstral.

URAC's Specialty Pharmacy Accreditation "provides an external validation of excellence in Specialty Pharmacy Management and provides Continuous Quality Improvement (CQI) oriented processes that improve operations and enhance compliance."

Regulation

Biologics or biological products for human use are regulated by the Center for Biologics Evaluation and Research (CBER), overseen by the Office of Medical Products and Tobacco, within the U.S. Food and Drug Administration which includes the Public Health Service Act and the Federal Food, Drug and Cosmetic Act. "CBER protects and advances the public health by ensuring that biological products are safe and effective and available to those who need them. CBER also provides the public with information to promote the safe and appropriate use of biological products."

Specialty Market Participants

There are multiple players in specialty drugs including the employer, the health plan, the pharmacy benefits manager and it is unclear who should be in charge of controlling costs and monitoring care. Pharmacies generally buy a product from a wholesaler and sell (Buy & Bill) it to the patient and provide basic drug use information and counseling. According to Maria Hardin, vice president of patient services for the National Organization for Rare Disorders (NORD), an alliance of voluntary health and patient advocacy groups working with rare diseases, "As the cost of drugs increases, management of the financial side has gotten more complex... The issues range from Medicare Part D to tiered benefits, prior authorizations, and no benefits. These patients need a pharmacy with the expertise and the clout to go to bat for them. If the patient doesn't get treated, the specialty pharmacy doesn't get paid."

Alexion Pharmaceuticals was one of the pioneers in the use of a business model of developing drugs to combat rare diseases. "Knowing the value of specialty drugs as well as its own stock is Alexion's business." Since other big pharmaceutical companies had tended to ignore these markets, Alexion had minimal competition at first. Insurance companies have generally been willing to pay high prices for such drugs; since few of their customers need the drugs, a high price does not significantly impact the insurance companies outlays. Alexion is thus seeking a stronger position in the lucrative rare disease market, and is willing to pay a premium to obtain that position. The rare disease market is seen as desirable because insurers have minimal motive to deny claims (due to small population sizes of patients) and are unable to negotiate better drug prices due to lack of competition. of May 2015, Alexion is currently seeking approval of its second drug, Strensiq. It will be used to treat Hypophosphatasia, a rare metabolic disorder. In 2015 Alexion estimated that Synageva, its specialty drug for Lysosomal Acid Lipase Deficiency, a fatal genetic disorder, could eventually have annual sales of more than $1 billion.

Companies like Magellan RX Management provide a "single source for high-touch patient care management to control side effects, patient support and education to ensure compliance or continued treatment, and specialized handling and distribution of medications directly to the patient or care provider. Specialty medications may be covered under either the medical or pharmacy benefit."

According to an article published in 2014 in the journal *Pharmacoeconomics*, "[s]pecialty pharmacies combine medication dispensing with clinical disease management. Their services have been used to improve patient outcomes and contain costs of specialty pharmaceuticals. These may be part of independent pharmacy businesses, retail pharmacy chains, wholesalers, pharmacy benefit managers (PBMs), or health insurance companies. Over the last several years, payers have been transitioning to obligate beneficiaries to receive self-administered agents (SAAs) from contracted specialty pharmacies, limiting the choice of acceptable specialty pharmacy providers (SPPs) for patient services."

Health Plans and PBMs

Managed Care Organizations (MCOs) contract with Specialty Pharmacy vendors. "Managed care organizations (MCOs) are using varied strategies to manage utilization and costs. For example, 58 % of 109 MCOs surveyed implement prior authorizations for MS specialty therapies." The Academy of Managed Care Pharmacy (AMCP) designates a product as a specialty drug if "[i]t requires a difficult or unusual process of delivery to the patient (preparation, handling, storage, inventory, distribution, Risk Evaluation and Mitigation Strategy (REMS) programs, data collection, or administration) or, Patient management prior to or following administration (monitoring, disease or therapeutic support systems)." Health plans consider 'high cost' (on average a minimum monthly costs of $US1,200) to be is a determining factor in identifying a specialty drug.

Independent Specialty Pharmacies

Tom Westrich, of St. Louis, Missouri-based Centric Health Resources, a specialty pharmacy, described how their specialty drugs treat ultra-orphan diseases with a total patient population of 20,000 nationwide.

Retail Pharmacies

Accredo is a specialty pharmacy that focuses on expensive, difficult to administer, "infused, injectable, and oral drugs" that are used recurrently to treat chronic and life-threatening diseases, and may cause adverse reactions, require temperature control or other specialized handling, yay have restrictions as determined by the FDA." Accredo "has access to many specialty medications with limited distribution, making us a recommended provider to patients, payors, and physicians."

The top ten specialty pharmacies in 2014 were CVS Specialty parent company CVS Health with $20.5B in sales, Express Scripts's Accredo at $15B, Walgreens Boots Alliance's Walgreens Specialty at $8.5B, UnitedHealth Group's OptumRx at $2.4B, Diplomat Pharmacy at $2.1B, Catamaran's BriovaRx at $2.0B, Specialty Prime Therapeutics's Prime Therapeutics at $1.8B, Omnicare's Advanced Care Scripts at $1.3B, Humana's RightsourceRx at $1.2B, Avella at $0.8B. All the other specialty pharmacies accounted for $22.4B of sales in 2014 with a total of $78B.

Hospitals and Physicians

In 2010 the United States enacted a new health law which had unintended consequences. Because of the 2010 law, drug companies like Genentech informed children's hospitals that they would no longer get discounts for certain cancer medicines such as the orphan drugs, Avastin, Herceptin, Rituxan, Tarceva, or Activase. This costs hospitals millions of dollars.

There is a debate about whether specialty drugs should be managed as a medical benefit or a pharmaceutical benefit. Infused or injected medications are usually covered under the medical benefit and oral medications are covered under the pharmacy benefit. Self-injected medications may be either. "Many biologics, such as chemotherapy drugs, are administered in a doctor's office and require extensive monitoring, further driving up costs." Chemotherapy is usually delivered intravenously, although a number of agents can be administered orally (e.g. specialty drugs, melphalan (trade name Alkeran), busulfan, capecitabine). Delcath Systems, Inc. (NASDAQ: DCTH) a specialty pharmaceutical and medical device company manufactures melphalan.

By 2011 the oral medications for cancer patients represented approximately 35% of cancer medications. Prior to the increase in cancer oral drugs community cancer centers were used to managing office-administered chemotherapy treatments. At that time "the majority of community oncology practices were unfamiliar with the process of prescribing and obtaining drugs that are covered under the pharmacy benefit" and "conventional retail pharmacy chains were ill-prepared to stock oral oncology agents, and were not set up to deliver the counseling that often accompanies these medications."

U.S. National Market Share

According to IMS Health "Specialty pharmaceutical spending is on the rise and is expected to increase from approximately $55 billion in 2005 to $1.7 trillion in 2030, according to the Pharmaceutical Care Management Association. That reflects an increase from 24% of total drug spend in 2005 to an estimated 44% of a health plan's total drug expenditure in 2030."

Mergers and Acquisitions Among Specialty Pharmacies

While CVS, Accredo, and Walgreens led the Specialty Pharmacies (SP) market in revenue in 2014, there are constant changes through mergers and acquisitions in terms of SPs and specialty distributors (SDs). The SP/SD network faces common strengths such as "in-depth clinical management, coordinated/comprehensive care, and early limited distribution network success" and common weaknesses, "lack of ability to customize services, poor integration experience and outcomes, and strained pharma relations." BioScrip was acquired by Walgreens in 2012. Specialty companies like Genzyme and MedImmune were acquired and are transitioning to a new business model.

Specialty Hubs

According to Nicolas Basta, by 2013 there was "a spate of new entities" called hub services, "mechanisms by which manufacturers can keep a grip on the marketplace" in specialty pharma. The "biggest and oldest of these organizations" are "offshoots of insurance companies or [Pharmacy benefit managers] PBMs, such as Express Scripts' combination of Accredo and CuraScript (both specialty pharmacies) and HealthBridge (physician and patient support). UnitedHealth, an insurance company, operates OptumRx, a PBM, which has a specialty unit within it. Cigna has Tel-Drug, a mail-order pharmacy and support system." Basta described how Hubs have been around since about 2002 "starting out as "reimbursement hubs," usually provided as a service by manufacturers to help patients and providers navigate the process of obtaining permission to use, and reimbursement for, expensive specialty therapies." Industry observers look to pioneering efforts by Genentech and Genzyme under the tenure of Henri Termeer, "when some of their earliest biotech products entered the marketplace." Specialty hubs provide reimbursement support to physicians and patients as well as patient education including medical hotlines. There is a voluntary program enrollment and registry intake with Patient Assistance Program management.

Affordability of Specialty Drugs and Patient Compliance with Care Plan

According to a 2007 study by employees of Express Scripts or its wholly owned subsidiary CuraScript on specialty pharmacy costs, if payers manage cost control through copayments with patients, there is an increased risk that patients will forego essential but expensive specialty drugs. and health outcomes were compromised. In 2007 these researchers suggested in the adoption of formularies and other traditional drug-management tools. They also recommended specialty drug utilization management programs that guide treatment plans and improve outpatient compliance.

Price Inflation Controversies

By 2010 Alexion Pharmaceuticals's Soliris, was considered to be the most expensive drug in the world.

In a 2012 article in the *New York Times,* journalist Andrew Pollack described how Don M. Bailey, a mechanical engineer by training who became interim president of Questcor Pharmaceuticals, Inc. (Questcor) in May 2007, initiated a new pricing model for Acthar in August 2007 when it was classified by FDA as an orphan drug and a specialty drug to treat infantile spasms. Questcor, a biopharmaceutical company, focuses on the treatment of patients with "serious, difficult-to-treat autoimmune and inflammatory disorders." Its primary product is FDA-approved Acthar, an injectable drug that is used for the treatment of 19 indications. At the same time Questcor created "an expanded safety net for patients using Acthar", provided a "group of Medical Science Liaisons to work with health care providers who are administering Acthar" and limited distribution to its sole specialty distributor, Curascript. The 2007 pricing model brought "Acthar in line with the cost of treatments for other very rare diseases." The cost for a course of treatment in 2007 was estimated at about "$80,000-$100,000." Acthar is now manufactured through a contractor on Prince Edward Island, Canada. The price increased from $40 a vial to $700 and continued to increase. By 2012 the price of a vial of Acthar was $28,400. and was considered to be one of the world's most expensive drugs in 2013.

By 2014 the price of Gilead's specialty drug for hepatitus C, Sovaldi or *sofosbuvir,* was $84,000 to $168,000 for a course of treatment in the U.S., £35,000 in the UK for 12 weeks. Sovaldi is on the World Health Organization's most important medications needed in a basic health system and the steep price is highly controversial. In 2014 the U.S. spent 742.6% more on hepatitis C medications than it did in 2013.

In September 2015, Martin Shkreli was criticized by several health organizations for obtaining manufacturing licenses on old, out-of-patent, life-saving medicines including pyrimethamine (brand name *Daraprim*), which is used to treat patients with toxoplasmosis, malaria, some cancers, and AIDS, and then increasing the price of the drug in the US from $13.50 to $750 per pill, a 5,455% increase. In an interview with *Bloomberg News*, Shkreli claimed that despite the price increase patient co-pays would actually be lower, that many patients would get the drug at no cost, that the company has expanded its free drug program, and that it sells half of the drugs for one dollar.

Captive Pharmacies

In 2015 Bloomberg News used the term 'captive pharmacies' to describe the alleged exclusive agreements such as that between the specialty mail-order pharmacy Philodor and Valeant, mail-order pharmacy Linden Care and Horizon Pharma Plc. In November 2015 Express Scripts Holding Co.—the largest U.S. manager of prescription drug benefits—"removed the mail-order pharmacy Linden Care LLC from its network after concluding it dispensed a large portion of its medications from Horizon Pharma Plc and didn't fulfill its contractual agreements." Express Scripts was "evaluating other 'captive pharmacies' that it said are mostly distributing Horizon drugs." In 2015 specialty pharmacies like "Philidor drew attention for the lengths they went to fill prescriptions with brand-name drugs and then secure insurance reimbursement.

Trans-pacific Partnership (TPP)

According to Pfenex, a clinical-stage biotechnology company, the proposed terms in the Trans-Pacific Partnership, a trade agreement between twelve Pacific Rim countries, would mean that all TPP partners would have to adopt the United States' lengthy drug patent exclusivity protection period of 12 years for biologics and specialty drugs.

Popular Culture

In 1981 an episode of the television series *Quincy, M.E.* starring star, Jack Klugman as Quincy, entitled "Seldom Silent, Never Heard" brought the plight of children with orphan diseases to public attention. In the episode Jeffrey, a young boy with Tourette's Syndrome, died after falling from a building. Dr. Arthur Ciotti (Michael Constantine), a medical doctor who had been researching Tourette's Syndrome for years wanted to study Jeffrey's brain to discover the cause and cure for the rare disease. He explained to Quincy that drug companies, like the one where he worked, were not interested in doing the research because so few people were afflicted with them that it was not financially viable. In 1982 another episode "Give Me Your Weak" Klugman as Quincy testified before Congress in an effort to get the Orphan Drug act passed. He was moved by the dilemma of a young mother with myoclonus.

References

- Biopolymer, Volume 8 Polyamides and Complex Proteinaceous Materials II, edited by S.R. Fahnestock & A. Steinbuchel, 2003 Wiley-VCH Verlag, pages 97-117 ISBN 978-3-527-30223-9

- Alan D. MacNaught, Andrew R. Wilkinson, ed. (1997). Compendium of Chemical Terminology: IUPAC Recommendations (the "Gold Book") (2nd ed.). Blackwell Science. ISBN 0865426848.

- Formulary Guidance (PDF), Medicare Modernization Act Final Guidelines - Formularies CMS Strategy for Affordable Access to Comprehensive Drug Coverage- Guidelines for Reviewing Prescription Drug Plan Formularies and Procedures, Baltimore, MD, 2004, retrieved 22 January 2016

- "The Growing Cost of Specialty Pharmacy-Is it Sustainable?". The American Journal of Managed Care. 18 February 2013. Retrieved 23 January 2016.

- Cortez, Michelle Fay; Lauerman, John (12 November 2015). "Valeant's Favorite Pharmacy Made Life Easy for Doctors, at a Price: Why Valeant Is an Unusual Pharmaceutical Ethics Case". Bloomberg.com. Retrieved 23 January 2016.

- "Just the Facts: Prescription Drug Formularies" (PDF), Cancer Action Network American Cancer Society (ACS-CAN), March 2014, retrieved 22 January 2016

- Schondelmeyer, Stephen W.; Purvis, Leigh (November 2015). "Trends in Retail Prices of Specialty Prescription Drugs Widely Used by Older Americans, 2006 to 2013" (PDF). AARP. Retrieved 23 January 2016.

- Tennille, Tracy (Feb 13, 2015). "First Genetically Modified Apple Approved for Sale in U.S.". Wall Street Journal. Retrieved Feb 2015.

- Feder, Barnaby J. (25 January 1988). "Kodak's Diversification Plan Moves Into a Higher Gear". New York Times. Retrieved 17 October 2015.

- Gleason, Alexander G. C.; Starner, C. I.; Ritter, S. T.; Van Houten, H. K.; Gunderson, B. W.; Shah, N. D. (September 2013). "Health plan utilization and costs of specialty drugs within 4 chronic conditions". Academy of Managed Care Pharmacy. 19 (7): 542–8. PMID 23964615. Retrieved 11 October 2015.

- Thomas, Kate; Pollack, Andrew (15 July 2015). "Specialty Pharmacies Proliferate, Along With Questions". Sinking Spring, Pa.: New York Times. Retrieved 5 October 2015.

- Kirchhoff, Suzanne M. (3 August 2015). "Specialty Drugs: Background and Policy Concerns" (PDF). Congressional Research Service. p. 26. Retrieved 27 October 2015.

- "Diplomat expands hepatitis C specialty services with acquisition of Burman's Specialty Pharmacy". PRNewswire. Flint, Michigan. 19 June 2015. Retrieved 5 October 2015.

- Kober, Scott (July 2008). "The Evolution of Specialty Pharmacy". Biotechnology Healthcare. pp. 50–51. PMC 2706163. Retrieved 3 November 2015.

- Langreth, Robert; Deprez, Esmé E. (29 October 2015). "The Tiny Pharmacy at the Center of Valeant's Money Mystery". Bloomberg. Retrieved 1 November 2015.

- Armstrong, Drew; Kitamura, Makiko (30 October 2015). "Valeant Says It's Cutting Ties With Troubled Pharmacy Philidor". Bloomberg. Retrieved 31 October 2015.

- "Valeant Pharmaceuticals International Inc cuts ties with Philidor as business practice controversy grows". Bloomberg News. 30 October 2015. Retrieved 1 November 2015.

- Lopez, Linette (23 October 2015). "The secret firm at the heart of Valeant's crisis has an alleged history of shady behavior with customers". Business Insider. Retrieved 1 November 2015.

- Rapoport, Michael (29 October 2015). "Valeant Countersues R&O Pharmacy Billing dispute draws attention to drug maker's work with specialty pharmacies". Retrieved 1 November 2015.

- "Valeant Pharmaceuticals Announces FDA Approval Of Jublia® for the Treatment of Onychomycosis". Valeant Pharmaceuticals. Laval, Quebec. 9 June 2014. Retrieved 1 November 2015.

6

Allied Fields of Biotechnology

Biotechnology is an interdisciplinary subject and spans across varied fields. This chapter covers agriculture, agricultural science, food industry and processing and medicine. The chapter provides a plethora of allied fields of biotechnology for a better comprehension of the respective subject.

Agriculture

Fields in Záhorie (Slovakia) – a typical Central European agricultural region

Agriculture is the cultivation of animals, plants and fungi for food, fiber, biofuel, medicinal plants and other products used to sustain and enhance human life. Agriculture was the key development in the rise of sedentary human civilization, whereby farming of domesticated species created food surpluses that nurtured the development of civilization. The study of agriculture is known as agricultural science. The history of agriculture dates back thousands of years, and its development has been driven and defined by greatly different climates, cultures, and technologies. Industrial agriculture based on large-scale monoculture farming has become the dominant agricultural methodology.

Modern agronomy, plant breeding, agrochemicals such as pesticides and fertilizers, and technological developments have in many cases sharply increased yields from cultivation, but at the same time have caused widespread ecological damage and negative human health effects. Selective breeding and modern practices in animal husbandry have similarly increased the output of meat, but have raised concerns about animal welfare and the health effects of the antibiotics, growth hormones, and other chemicals commonly used in industrial meat production. Genetically modified organisms are an increasing component of agriculture, although they are banned in several countries. Agricultural food production and water management are increasingly becoming global issues that are fostering debate on a number of fronts. Significant degradation of land and water resources, including the depletion of aquifers, has been observed in recent decades, and the effects of global warming on agriculture and of agriculture on global warming are still not fully understood.

Domestic sheep and a cow (heifer) pastured together in South Africa

The major agricultural products can be broadly grouped into foods, fibers, fuels, and raw materials. Specific foods include cereals (grains), vegetables, fruits, oils, meats and spices. Fibers include cotton, wool, hemp, silk and flax. Raw materials include lumber and bamboo. Other useful materials are also produced by plants, such as resins, dyes, drugs, perfumes, biofuels and ornamental products such as cut flowers and nursery plants. Over one third of the world's workers are employed in agriculture, second only to the service sector, although the percentages of agricultural workers in developed countries has decreased significantly over the past several centuries.

Etymology and Terminology

The word *agriculture* is a late Middle English adaptation of Latin *agricultūra*, from *ager*, "field", and *cultūra*, "cultivation" or "growing". Agriculture usually refers to human activities, although it is also observed in certain species of ant, termite and ambrosia beetle. To practice agriculture means to use natural resources to "produce commodities which maintain life, including food, fiber, forest products, horticultural crops, and their related services." This definition includes arable farming or agronomy, and horticulture, all terms for the growing of plants, animal husbandry and forestry. A distinction is sometimes made between forestry and agriculture, based on the former's longer management rotations, extensive versus intensive management practices and development mainly by nature, rather than by man. Even then, it is acknowledged that there is a large amount of knowledge transfer and overlap between silviculture (the management of forests) and agriculture. In traditional farming, the two are often combined even on small landholdings, leading to the term agroforestry.

History

Agriculture began independently in different parts of the globe, and included a diverse range of taxa. At least 11 separate regions of the Old and New World were involved as independent centers of origin.

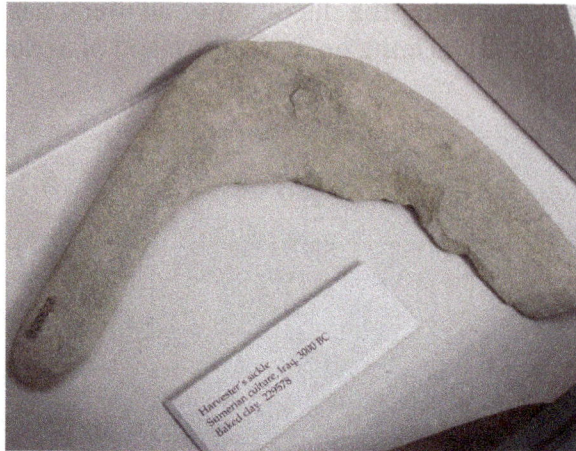

A Sumerian harvester's sickle made from baked clay (c. 3000 BC)

Wild grains were collected and eaten from at least 20,000 BC. From around 9500 BC, the eight Neolithic founder crops, emmer and einkorn wheat, hulled barley, peas, lentils, bitter vetch, chick peas and flax were cultivated in the Levant. Rice was domesticated in China between 8,200 and 13,500 years ago, followed by mung, soy and azuki beans. Pigs were domesticated in Mesopotamia around 15,000 years ago. Cattle were domesticated from the wild aurochs in the areas of modern Turkey and Pakistan some 10,500 years ago. Sheep were domesticated in Mesopotamia between 11,000 and 9,000 BC. Sugarcane and some root vegetables were domesticated in New Guinea around 7,000 BC. Sorghum was domesticated in the Sahel region of Africa by 5000 BC. In the Andes of South America, the potato was domesticated between 8,000 and 5,000 BC, along with beans, coca, llamas, alpacas, and guinea pigs. Cotton was domesticated in Peru by 3,600 BC, and was independently domesticated in Eurasia at an unknown time. In Mesoamerica, wild teosinte was domesticated to maize by 4,000 BC.

In the Middle Ages, both in the Islamic world and in Europe, agriculture was transformed with improved techniques and the diffusion of crop plants, including the introduction of sugar, rice, cotton and fruit trees such as the orange to Europe by way of Al-Andalus. After 1492, the Columbian exchange brought New World crops such as maize, potatoes, sweet potatoes and manioc to Europe, and Old World crops such as wheat, barley, rice and turnips, and livestock including horses, cattle, sheep and goats to the Americas. Irrigation, crop rotation, and fertilizers were introduced soon after the Neolithic Revolution and developed much further in the past 200 years, starting with the British Agricultural Revolution. Since 1900, agriculture in the developed nations, and to a lesser extent in the developing world, has seen large rises in productivity as human labor has been replaced by mechanization, and assisted by synthetic fertilizers, pesticides, and selective breeding. The Haber-Bosch method allowed the synthesis of ammonium nitrate fertilizer on an industrial scale, greatly increasing crop yields. Modern agriculture has raised political issues including water pollution, biofuels, genetically modified organisms, tariffs and farm subsidies, leading to alternative approaches such as the organic movement.

Agriculture and Civilization

Civilization was the product of the Agricultural Neolithic Revolution. In the course of history, civilization coincided in space with fertile areas (The Fertile Crescent) and most intensive state formation took place in circumscribed agricultural lands (Carneiro's circumscription theory). The Great Wall of China and the Roman limes demarcated the same northern frontier of the basic (cereal) agriculture. This cereal belt nourished the belt of great civilizations formed in the Axial Age and connected by the famous Silk Road.

Ancient Egyptians, whose agriculture depended exclusively on the Nile, deified the river, worshipped, and exalted it in a great hymn. The Chinese imperial court issued numerous edicts, stating: "Agriculture is the foundation of this Empire." Egyptian, Mesopotamian, Chinese, and Inca Emperors themselves plowed ceremonial fields in order to show personal example to everyone.

Ancient strategists, Chinese Guan Zhong and Shang Yang and Indian Kautilya, drew doctrines linking agriculture with military power. Agriculture defined the limits on how large and for how long an army could be mobilized. Shang Yang called agriculture and war the *One*. In the vast human pantheon of agricultural deities there are several deities who combined the functions of agriculture and war.

As the Neolithic Agricultural Revolution produced civilization, the modern Agricultural Revolution, begun in Britain (British Agricultural Revolution), made possible the Industrial civilization. The first precondition for industry was greater yields by less manpower, resulting in greater percentage of manpower available for non-agricultural sectors.

Types of Agriculture

Reindeer herds form the basis of pastoral agriculture for several Arctic and Subarctic peoples.

Pastoralism involves managing domesticated animals. In nomadic pastoralism, herds of livestock are moved from place to place in search of pasture, fodder, and water. This type of farming is practised in arid and semi-arid regions of Sahara, Central Asia and some parts of India like Rajasthan and Jammu and Kashmir.

In shifting cultivation, a small area of a forest is cleared by cutting down all the trees and the area is burned. Now that area is used for growing crops after some years when the area become less fertile then it is abandoned. Again a patch is selected and this process is performed. This type of farming is practiced mainly in areas with abundant rainfall, so forest regenerate very quickly. They are practiced in North-East India, South-East Asia and the Amazon Basin.

Subsistence farming is practiced to satisfy family or local needs alone, with little left over for transport elsewhere. It is intensively practiced in Monsoon Asia and South-East Asia.

In intensive farming, the crops are cultivated for commercial purpose i.e., for selling. The main motive of the farmer is to make profit. This type of farming is mainly practice in western countries like Canada and USA.

Contemporary Agriculture

Satellite image of farming in Minnesota

In the past century, agriculture has been characterized by increased productivity, the substitution of synthetic fertilizers and pesticides for labor, water pollution, and farm subsidies. In recent years there has been a backlash against the external environmental effects of conventional agriculture, resulting in the organic and sustainable agriculture movements. One of the major forces behind this movement has been the European Union, which first certified organic food in 1991 and began reform of its Common Agricultural Policy (CAP) in 2005 to phase out commodity-linked farm subsidies, also known as decoupling. The growth of organic farming has renewed research in alternative technologies such as integrated pest management and selective breeding. Recent mainstream technological developments include genetically modified food.

In 2007, higher incentives for farmers to grow non-food biofuel crops combined with other factors, such as over development of former farm lands, rising transportation costs, climate change, growing consumer demand in China and India, and population growth, caused food shortages in Asia, the Middle East, Africa, and Mexico, as well as rising food prices around the globe. As of December 2007, 37 countries faced food crises, and 20 had imposed some sort of food-price controls. Some of these shortages resulted in food riots and even deadly stampedes. The International Fund for Agricultural Development posits that an increase in smallholder agriculture may be part of the solution to concerns about food prices and overall food security. They in part base this on the experience of Vietnam, which went from a food importer to large food exporter and saw a significant drop in poverty, due mainly to the development of smallholder agriculture in the country.

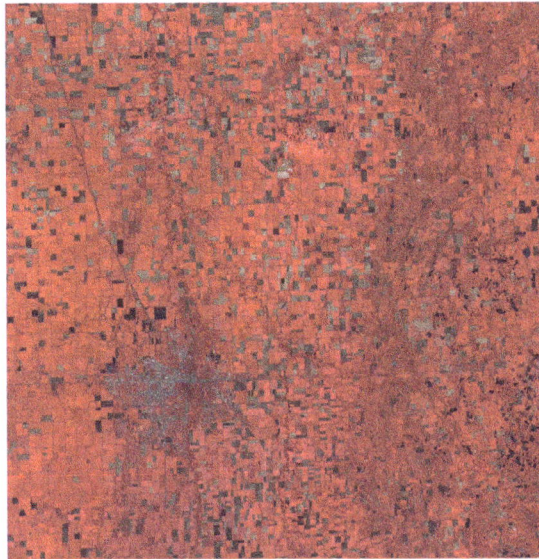

Infrared image of the above farms. Various colors indicate healthy crops (red), flooding (black) and unwanted pesticides (brown).

Disease and land degradation are two of the major concerns in agriculture today. For example, an epidemic of stem rust on wheat caused by the Ug99 lineage is currently spreading across Africa and into Asia and is causing major concerns due to crop losses of 70% or more under some conditions. Approximately 40% of the world's agricultural land is seriously degraded. In Africa, if current trends of soil degradation continue, the continent might be able to feed just 25% of its population by 2025, according to United Nations University's Ghana-based Institute for Natural Resources in Africa.

Agrarian structure is a long-term structure in the Braudelian understanding of the concept. On a larger scale the agrarian structure is more dependent on the regional, social, cultural and historical factors than on the state's undertaken activities. Like in Poland, where despite running an intense agrarian policy for many years, the agrarian structure in 2002 has much in common with that found in 1921 soon after the partitions period.

In 2009, the agricultural output of China was the largest in the world, followed by the European Union, India and the United States, according to the International Monetary Fund. Economists measure the total factor productivity of agriculture and by this measure agriculture in the United States is roughly 1.7 times more productive than it was in 1948.

Workforce

As of 2011, the International Labour Organization states that approximately one billion people, or over 1/3 of the available work force, are employed in the global agricultural sector. Agriculture constitutes approximately 70% of the global employment of children, and in many countries employs the largest percentage of women of any industry. The service sector only overtook the agricultural sector as the largest global employer in 2007. Between 1997 and 2007, the percentage of people employed in agriculture fell by over four percentage points, a trend that is expected to continue. The number of people employed in agriculture varies widely on a per-country basis, ranging from less than 2% in countries like the US and Canada to over 80% in many African nations. In developed countries, these figures are significantly lower than in previous centuries. During the 16th century in Europe, for example, between 55 and 75 percent of the population was engaged in agriculture, depending on the country. By the 19th century in Europe, this had dropped to between 35 and 65 percent. In the same countries today, the figure is less than 10%.

Safety

Rollover protection bar on a Fordson tractor

Agriculture, specifically farming, remains a hazardous industry, and farmers worldwide remain at high risk of work-related injuries, lung disease, noise-induced hearing loss, skin diseases, as well as certain cancers related to chemical use and prolonged sun exposure. On industrialized farms, injuries frequently involve the use of agricultural machinery, and a common cause of fatal agricultural injuries in developed countries is tractor rollovers. Pesticides and other chemicals used in farming can also be hazardous to worker health, and workers exposed to pesticides may experience illness or have children with birth defects. As an industry in which families commonly share in work and live on the farm itself, entire families can be at risk for injuries, illness, and death. Common causes of fatal injuries among young farm workers include drowning, machinery and motor vehicle-related accidents.

The International Labour Organization considers agriculture "one of the most hazardous of all economic sectors." It estimates that the annual work-related death toll among agricultural employees is at least 170,000, twice the average rate of other jobs. In addition, incidences of death, injury and illness related to agricultural activities often go unreported. The organization has developed the Safety and Health in Agriculture Convention, 2001, which covers the range of risks in the agriculture occupation, the prevention of these risks and the role that individuals and organizations engaged in agriculture should play.

Agricultural Production Systems

Crop Cultivation Systems

Cropping systems vary among farms depending on the available resources and constraints; geography and climate of the farm; government policy; economic, social and political pressures; and the philosophy and culture of the farmer.

Rice cultivation at a paddy field in Bihar state of India

Shifting cultivation (or slash and burn) is a system in which forests are burnt, releasing nutrients to support cultivation of annual and then perennial crops for a period of several years. Then the plot is left fallow to regrow forest, and the farmer moves to a new plot, returning after many more years (10 – 20). This fallow period is shortened if population density grows, requiring the input of nutrients (fertilizer or manure) and some manual pest control. Annual cultivation is the next phase of intensity in which there is no fallow period. This requires even greater nutrient and pest control inputs.

The Banaue Rice Terraces in Ifugao, Philippines

Further industrialization led to the use of monocultures, when one cultivar is planted on a large acreage. Because of the low biodiversity, nutrient use is uniform and pests tend to build up, necessitating the greater use of pesticides and fertilizers. Multiple cropping, in which several crops are grown sequentially in one year, and intercropping, when several crops are grown at the same time, are other kinds of annual cropping systems known as polycultures.

In subtropical and arid environments, the timing and extent of agriculture may be limited by rainfall, either not allowing multiple annual crops in a year, or requiring irrigation. In all of these environments perennial crops are grown (coffee, chocolate) and systems are practiced such as agroforestry. In temperate environments, where ecosystems were predominantly grassland or prairie, highly productive annual farming is the dominant agricultural system.

Crop Statistics

Important categories of crops include cereals and pseudocereals, pulses (legumes), forage, and fruits and vegetables. Specific crops are cultivated in distinct growing regions throughout the world. In millions of metric tons, based on FAO estimate.

Top agricultural products, by crop types (million tonnes) 2004 data	
Cereals	2,263
Vegetables and melons	866
Roots and tubers	715
Milk	619
Fruit	503
Meat	259
Oilcrops	133
Fish (2001 estimate)	130
Eggs	63
Pulses	60
Vegetable fiber	30
Source: Food and Agriculture Organization (FAO)	

Top agricultural products, by individual crops (million tonnes) 2011 data	
Sugar cane	1794
Maize	883
Rice	722
Wheat	704
Potatoes	374
Sugar beet	271
Soybeans	260
Cassava	252
Tomatoes	159
Barley	134
Source: Food and Agriculture Organization (FAO)	

Livestock Production Systems

Animals, including horses, mules, oxen, water buffalo, camels, llamas, alpacas, donkeys, and dogs, are often used to help cultivate fields, harvest crops, wrangle other animals, and transport farm products to buyers. Animal husbandry not only refers to the breeding and raising of animals for meat or to harvest animal products (like milk, eggs, or wool) on a continual basis, but also to the breeding and care of species for work and companionship.

Livestock production systems can be defined based on feed source, as grassland-based, mixed, and landless. As of 2010, 30% of Earth's ice- and water-free area was used for producing livestock, with the sector employing approximately 1.3 billion people. Between the 1960s and the 2000s, there was a significant increase in livestock production, both by numbers and by carcass weight, especially among beef, pigs and chickens, the latter of which had production increased by almost a factor of 10. Non-meat animals, such as milk cows and egg-producing chickens, also showed significant production increases. Global cattle, sheep and goat populations are expected to continue to increase sharply through 2050. Aquaculture or fish farming, the production of fish for human consumption in confined operations, is one of the fastest growing sectors of food production, growing at an average of 9% a year between 1975 and 2007.

Ploughing rice paddy fields with water buffalo, in Indonesia

Oxen driven ploughs in India

During the second half of the 20th century, producers using selective breeding focused on creating livestock breeds and crossbreeds that increased production, while mostly disregarding the need to preserve genetic diversity. This trend has led to a significant decrease in genetic diversity and resources among livestock breeds, leading to a corresponding decrease in disease resistance and local adaptations previously found among traditional breeds.

Grassland based livestock production relies upon plant material such as shrubland, rangeland, and pastures for feeding ruminant animals. Outside nutrient inputs may be used, however manure is returned directly to the grassland as a major nutrient source. This system is particularly important in areas where crop production is not feasible because of climate or soil, representing 30 – 40 million pastoralists. Mixed production systems use grassland, fodder crops and grain feed crops as feed for ruminant and monogastric (one stomach; mainly chickens and pigs) livestock. Manure is typically recycled in mixed systems as a fertilizer for crops.

Landless systems rely upon feed from outside the farm, representing the de-linking of crop and livestock production found more prevalently in Organisation for Economic Co-operation and Development(OECD) member countries. Synthetic fertilizers are more heavily relied upon for crop production and manure utilization becomes a challenge as well as a source for pollution. Industrialized countries use these operations to produce much of the global supplies of poultry and pork. Scientists estimate that 75% of the growth in livestock production between 2003 and 2030 will be in confined animal feeding operations, sometimes called factory farming. Much of this growth is happening in developing countries in Asia, with much smaller amounts of growth in Africa. Some of the practices used in commercial livestock production, including the usage of growth hormones, are controversial.

Production Practices

Road leading across the farm allows machinery access to the farm for production practices

Farming is the practice of agriculture by specialized labor in an area primarily devoted to agricultural processes, in service of a dislocated population usually in a city.

Tillage is the practice of plowing soil to prepare for planting or for nutrient incorporation or for pest control. Tillage varies in intensity from conventional to no-till. It may improve productivity by warming the soil, incorporating fertilizer and controlling weeds, but also renders soil more prone to erosion, triggers the decomposition of organic matter releasing CO_2, and reduces the abundance and diversity of soil organisms.

Pest control includes the management of weeds, insects, mites, and diseases. Chemical (pesticides), biological (biocontrol), mechanical (tillage), and cultural practices are used. Cultural practices include crop rotation, culling, cover crops, intercropping, composting, avoidance, and resistance. Integrated pest management attempts to use all of these methods to keep pest populations below the number which would cause economic loss, and recommends pesticides as a last resort.

Nutrient management includes both the source of nutrient inputs for crop and livestock production, and the method of utilization of manure produced by livestock. Nutrient inputs can be chemical inorganic fertilizers, manure, green manure, compost and mined minerals. Crop nutrient use may also be managed using cultural techniques such as crop rotation or a fallow period. Manure is used either by holding livestock where the feed crop is growing, such as in managed intensive rotational grazing, or by spreading either dry or liquid formulations of manure on cropland or pastures.

Water management is needed where rainfall is insufficient or variable, which occurs to some degree in most regions of the world. Some farmers use irrigation to supplement rainfall. In other areas such as the Great Plains in the U.S. and Canada, farmers use a fallow year to conserve soil moisture to use for growing a crop in the following year. Agriculture represents 70% of freshwater use worldwide.

According to a report by the International Food Policy Research Institute, agricultural technologies will have the greatest impact on food production if adopted in combination with each other; using a model that assessed how eleven technologies could impact agricultural productivity, food security and trade by 2050, the International Food Policy Research Institute found that the number of people at risk from hunger could be reduced by as much as 40% and food prices could be reduced by almost half.

"Payment for ecosystem services (PES) can further incentivise efforts to green the agriculture sector. This is an approach that verifies values and rewards the benefits of ecosystem services provided by green agricultural practices." "Innovative PES measures could include reforestation payments made by cities to upstream communities in rural areas of shared watersheds for improved quantities and quality of fresh water for municipal users. Ecoservice payments by farmers to upstream forest stewards for properly managing the flow of soil nutrients, and methods to monetise the carbon sequestration and emission reduction credit benefits of green agriculture practices in order to compensate farmers for their efforts to restore and build SOM and employ other practices."

Crop Alteration and Biotechnology

Crop alteration has been practiced by humankind for thousands of years, since the beginning of civilization. Altering crops through breeding practices changes the genetic make-up of a plant to develop crops with more beneficial characteristics for humans, for example, larger fruits or seeds, drought-tolerance, or resistance to pests. Significant advances in plant breeding ensued after the work of geneticist Gregor Mendel. His work on dominant and recessive alleles, although initially largely ignored for almost 50 years, gave plant breeders a better understanding of genetics and breeding techniques. Crop breeding includes techniques such as plant selection with desirable traits, self-pollination and cross-pollination, and molecular techniques that genetically modify the organism.

Domestication of plants has, over the centuries increased yield, improved disease resistance and drought tolerance, eased harvest and improved the taste and nutritional value of crop plants. Careful selection and breeding have had enormous effects on the characteristics of crop plants. Plant selection and breeding in the 1920s and 1930s improved pasture (grasses and clover) in New Zealand. Extensive X-ray and ultraviolet induced mutagenesis efforts (i.e. primitive genetic engineering) during the 1950s produced the modern commercial varieties of grains such as wheat, corn (maize) and barley.

Tractor and chaser bin

The Green Revolution popularized the use of conventional hybridization to sharply increase yield by creating "high-yielding varieties". For example, average yields of corn (maize) in the USA have increased from around 2.5 tons per hectare (t/ha) (40 bushels per acre) in 1900 to about 9.4 t/ha (150 bushels per acre) in 2001. Similarly, worldwide average wheat yields have increased from less than 1 t/ha in 1900 to more than 2.5 t/ha in 1990. South American average wheat yields are around 2 t/ha, African under 1 t/ha, and Egypt and Arabia up to 3.5 to 4 t/ha with irrigation. In contrast, the average wheat yield in countries such as France is over 8 t/ha. Variations in yields are due mainly to variation in climate, genetics, and the level of intensive farming techniques (use of fertilizers, chemical pest control, growth control to avoid lodging).

Genetic Engineering

Genetically modified organisms (GMO) are organisms whose genetic material has been altered by genetic engineering techniques generally known as recombinant DNA technology. Genetic engineering has expanded the genes available to breeders to utilize in creating desired germlines for new crops. Increased durability, nutritional content, insect and virus resistance and herbicide tolerance are a few of the attributes bred into crops through genetic engineering. For some, GMO crops cause food safety and food labeling concerns. Numerous countries have placed restrictions on the production, import or use of GMO foods and crops, which have been put in place due to concerns over potential health issues, declining agricultural diversity and contamination of non-GMO crops. Currently a global treaty, the Biosafety Protocol, regulates the trade of GMOs. There is ongoing discussion regarding the labeling of foods made from GMOs, and while the EU currently requires all GMO foods to be labeled, the US does not.

Herbicide-resistant seed has a gene implanted into its genome that allows the plants to tolerate exposure to herbicides, including glyphosates. These seeds allow the farmer to grow a crop that can be sprayed with herbicides to control weeds without harming the resistant crop. Herbicide-tolerant crops are used by farmers worldwide. With the increasing use of herbicide-tolerant crops, comes an increase in the use of glyphosate-based herbicide sprays. In some areas glyphosate resistant weeds have developed, causing farmers to switch to other herbicides. Some studies also link widespread glyphosate usage to iron deficiencies in some crops, which is both a crop production and a nutritional quality concern, with potential economic and health implications.

Other GMO crops used by growers include insect-resistant crops, which have a gene from the soil bacterium *Bacillus thuringiensis* (Bt), which produces a toxin specific to insects. These crops protect plants from damage by insects. Some believe that similar or better pest-resistance traits can be acquired through traditional breeding practices, and resistance to various pests can be gained through hybridization or cross-pollination with wild species. In some cases, wild species are the primary source of resistance traits; some tomato cultivars that have gained resistance to at least 19 diseases did so through crossing with wild populations of tomatoes.

Environmental Impact

Water pollution in a rural stream due to runoff from farming activity in New Zealand

Agriculture, as implemented through the method of farming, imposes external costs upon society through pesticides, nutrient runoff, excessive water usage, loss of natural environment and assorted other problems. A 2000 assessment of agriculture in the UK determined total external costs for 1996 of £2,343 million, or £208 per hectare. A 2005 analysis of these costs in the USA concluded that cropland imposes approximately $5 to 16 billion ($30 to $96 per hectare), while livestock production imposes $714 million. Both studies, which focused solely on the fiscal impacts, concluded that more should be done to internalize external costs. Neither included subsidies in their analysis, but they noted that subsidies also influence the cost of agriculture to society. In 2010, the International Resource Panel of the United Nations Environment Programme published a report assessing the environmental impacts of consumption and production. The study found that agriculture and food consumption are two of the most important drivers of environmental pressures, particularly habitat change, climate change, water use and toxic emissions. The 2011 UNEP Green Economy report states that "[a]gricultural operations, excluding land use changes, produce approximately 13 per cent of anthropogenic global GHG emissions. This includes GHGs emitted by the use of inorganic fertilisers agro-chemical pesticides and herbicides; (GHG emissions resulting from production of these inputs are included in industrial emissions); and fossil fuel-energy inputs. "On average we find that the total amount of fresh residues from agricultural and forestry production for second- generation biofuel production amounts to 3.8 billion tonnes per year between 2011 and 2050 (with an average annual growth rate of 11 per cent throughout the period analysed, accounting for higher growth during early years, 48 per cent for 2011–2020 and an average 2 per cent annual expansion after 2020)."

Livestock Issues

A senior UN official and co-author of a UN report detailing this problem, Henning Steinfeld, said "Livestock are one of the most significant contributors to today's most serious environmental problems". Livestock production occupies 70% of all land used for agriculture, or 30% of the land surface of the planet. It is one of the largest sources of greenhouse gases, responsible for 18% of the world's greenhouse gas emissions as measured in CO_2 equivalents. By comparison, all transportation emits 13.5% of the CO_2. It produces 65% of human-related nitrous oxide (which has 296 times the global warming potential of CO_2) and 37% of all human-induced methane (which is 23 times as warming as CO_2.) It also generates 64% of the ammonia emission. Livestock expansion is cited as a key factor driving deforestation; in the Amazon basin 70% of previously forested area is now occupied by pastures and the remainder used for feedcrops. Through deforestation and land degradation, livestock is also driving reductions in biodiversity. Furthermore, the UNEP states that "methane emissions from global livestock are projected to increase by 60 per cent by 2030 under current practices and consumption patterns."

Land and Water Issues

Land transformation, the use of land to yield goods and services, is the most substantial way humans alter the Earth's ecosystems, and is considered the driving force in the loss of biodiversity. Estimates of the amount of land transformed by humans vary from 39 to 50%. Land degradation, the long-term decline in ecosystem function and productivity, is estimated to be occurring on 24% of land worldwide, with cropland overrepresented. The UN-FAO report cites land management as the driving factor behind degradation and reports that 1.5 billion people rely upon the degrading land. Degradation can be deforestation, desertification, soil erosion, mineral depletion, or chemical degradation (acidification and salinization).

Eutrophication, excessive nutrients in aquatic ecosystems resulting in algal blooms and anoxia, leads to fish kills, loss of biodiversity, and renders water unfit for drinking and other industrial uses. Excessive fertilization and manure application to cropland, as well as high livestock stocking densities cause nutrient (mainly nitrogen and phosphorus) runoff and leaching from agricultural land. These nutrients are major nonpoint pollutants contributing to eutrophication of aquatic ecosystems.

Agriculture accounts for 70 percent of withdrawals of freshwater resources. Agriculture is a major draw on water from aquifers, and currently draws from those underground water sources at an unsustainable rate. It is long known that aquifers in areas as diverse as northern China, the Upper Ganges and the western US are being depleted, and new research extends these problems to aquifers in Iran, Mexico and Saudi Arabia. Increasing pressure is being placed on water resources by industry and urban areas, meaning that water scarcity is increasing and agriculture is facing the challenge of producing more food for the world's growing population with reduced water resources. Agricultural water usage can also cause major environmental problems, including the destruction of natural wetlands, the spread of water-borne diseases, and land degradation through salinization and waterlogging, when irrigation is performed incorrectly.

Pesticides

Pesticide use has increased since 1950 to 2.5 million short tons annually worldwide, yet crop loss from pests has remained relatively constant. The World Health Organization estimated in 1992 that 3 million pesticide poisonings occur annually, causing 220,000 deaths. Pesticides select for pesticide resistance in the pest population, leading to a condition termed the "pesticide treadmill" in which pest resistance warrants the development of a new pesticide.

An alternative argument is that the way to "save the environment" and prevent famine is by using pesticides and intensive high yield farming, a view exemplified by a quote heading the Center for Global Food Issues website: 'Growing more per acre leaves more land for nature'. However, critics argue that a trade-off between the environment and a need for food is not inevitable, and that pesticides simply replace good agronomic practices such as crop rotation. The UNEP introduces the Push–pull agricultural pest management technique which involves intercropping that uses plant aromas to repel or push away pests while pulling in or attracting the right insects. "The implementation of push-pull in eastern Africa has significantly increased maize yields and the combined cultivation of N-fixing forage crops has enriched the soil and has also provided farmers with feed for livestock. With increased livestock operations, the farmers are able to produce meat, milk and other dairy products and they use the manure as organic fertiliser that returns nutrients to the fields."

Climate Change

Climate change has the potential to affect agriculture through changes in temperature, rainfall (timing and quantity), CO_2, solar radiation and the interaction of these elements. Extreme events, such as droughts and floods, are forecast to increase as climate change takes hold. Agriculture is among sectors most vulnerable to the impacts of climate change; water supply for example, will be critical to sustain agricultural production and provide the increase in food output required to sustain the world's growing population. Fluctuations in the flow of rivers are likely to increase in the twenty-first century. Based on the experience of countries in the Nile river basin (Ethiopia, Kenya and Sudan) and other developing countries, depletion of water resources during seasons crucial for agriculture can lead to a decline in yield by up to 50%. Transformational approaches will be needed to manage natural resources in the future. For example, policies, practices and tools promoting climate-smart agriculture will be important, as will better use of scientific information on climate for assessing risks and vulnerability. Planners and policy-makers will need to help create suitable policies that encourage funding for such agricultural transformation.

Agriculture in its many forms can both mitigate or worsen global warming. Some of the increase in CO_2 in the atmosphere comes from the decomposition of organic matter in the soil, and much of the methane emitted into the atmosphere is caused by the decomposition of organic matter in wet soils such as rice paddy fields, as well as the normal digestive activities of farm animals. Further, wet or anaerobic soils also lose nitrogen through denitrification, releasing the greenhouse gases nitric oxide and nitrous oxide. Changes in management can reduce the release of these greenhouse gases, and soil can further be used to sequester some of the CO_2 in the atmosphere. Informed by the UNEP, "[a]griculture also produces about 58 per cent of global nitrous oxide emissions and about 47 per cent of global methane emissions. Cattle and rice farms release methane, fertilized fields release nitrous oxide, and the cutting down of rainforests to grow crops or raise livestock releases carbon dioxide. Both of these gases have a far greater global warming potential per tonne than CO_2 (298 times and 25 times respectively)."

There are several factors within the field of agriculture that contribute to the large amount of CO_2 emissions. The diversity of the sources ranges from the production of farming tools to the transport of harvested produce. Approximately 8% of the national carbon footprint is due to agricultural sources. Of that, 75% is of the carbon emissions released from the production of crop assisting chemicals. Factories producing insecticides, herbicides, fungicides, and fertilizers are a major culprit of the greenhouse gas. Productivity on the farm itself and the use of machinery is another source of the carbon emission. Almost all the industrial machines used in modern farming are powered by fossil fuels. These instruments are burning fossil fuels from the beginning of the process to the end. Tractors are the root of this source. The tractor is going to burn fuel and release CO_2 just to run. The amount of emissions from the machinery increase with the attachment of different units and need for more power. During the soil preparation stage tillers and plows will be used to disrupt the soil. During growth watering pumps and sprayers are used to keep the crops hydrated. And when the crops are ready for picking a forage or combine harvester is used. These types of machinery all require additional energy which leads to increased carbon dioxide emissions from the basic tractors. The final major contribution to CO_2 emissions in agriculture is in the final transport of produce. Local farming suffered a decline over the past century due to large amounts of farm subsidies. The majority of crops are shipped hundreds of miles to various processing plants before ending up in the grocery store. These shipments are made using fossil fuel burning modes of transportation. Inevitably these transport adds to carbon dioxide emissions.

Sustainability

Some major organizations are hailing farming within agroecosystems as the way forward for mainstream agriculture. Current farming methods have resulted in over-stretched water resources, high levels of erosion and reduced soil fertility. According to a report by the International Water Management Institute and UNEP, there is not enough water to continue farming using current practices; therefore how critical water, land, and ecosystem resources are used to boost crop yields must be reconsidered. The report suggested assigning value to ecosystems, recognizing environmental and livelihood tradeoffs, and balancing the rights of a variety of users and interests. Inequities that result when such measures are adopted would need to be addressed, such as the reallocation of water from poor to rich, the clearing of land to make way for more productive farmland, or the preservation of a wetland system that limits fishing rights.

Technological advancements help provide farmers with tools and resources to make farming more sustainable. New technologies have given rise to innovations like conservation tillage, a farming process which helps prevent land loss to erosion, water pollution and enhances carbon sequestration.

According to a report by the International Food Policy Research Institute (IFPRI), agricultural technologies will have the greatest impact on food production if adopted in combination with each other; using a model that assessed how eleven technologies could impact agricultural productivity, food security and trade by 2050, IFPRI found that the number of people at risk from hunger could be reduced by as much as 40% and food prices could be reduced by almost half.

Agricultural Economics

Agricultural economics refers to economics as it relates to the "production, distribution and consumption of [agricultural] goods and services". Combining agricultural production with general theories of marketing and business as a discipline of study began in the late 1800s, and grew significantly through the 20th century. Although the study of agricultural economics is relatively recent, major trends in agriculture have significantly affected national and international economies throughout history, ranging from tenant farmers and sharecropping in the post-American Civil War Southern United States to the European feudal system of manorialism. In the United States, and elsewhere, food costs attributed to food processing, distribution, and agricultural marketing, sometimes referred to as the value chain, have risen while the costs attributed to farming have declined. This is related to the greater efficiency of farming, combined with the increased level of value addition (e.g. more highly processed products) provided by the supply chain. Market concentration has increased in the sector as well, and although the total effect of the increased market concentration is likely increased efficiency, the changes redistribute economic surplus from producers (farmers) and consumers, and may have negative implications for rural communities.

National government policies can significantly change the economic marketplace for agricultural products, in the form of taxation, subsidies, tariffs and other measures. Since at least the 1960s, a combination of import/export restrictions, exchange rate policies and subsidies have affected farmers in both the developing and developed world. In the 1980s, it was clear that non-subsidized farmers in developing countries were experiencing adverse effects from national policies that created artificially low global prices for farm products. Between the mid-1980s and the early 2000s, several international agreements were put into place that limited agricultural tariffs, subsidies and other trade restrictions.

However, as of 2009, there was still a significant amount of policy-driven distortion in global agricultural product prices. The three agricultural products with the greatest amount of trade distortion were sugar, milk and rice, mainly due to taxation. Among the oilseeds, sesame had the greatest amount of taxation, but overall, feed grains and oilseeds had much lower levels of taxation than livestock products. Since the 1980s, policy-driven distortions have seen a greater decrease among livestock products than crops during the worldwide reforms in agricultural policy. Despite this progress, certain crops, such as cotton, still see subsidies in developed countries artificially deflating global prices, causing hardship in developing countries with non-subsidized farmers. Unprocessed commodities (i.e. corn, soybeans, cows) are generally graded to indicate quality. The quality affects the price the producer receives. Commodities are generally reported by production quantities, such as volume, number or weight.

Agricultural Science

Agricultural science is a broad multidisciplinary field of biology that encompasses the parts of exact, natural, economic and social sciences that are used in the practice and understanding of agriculture. (Veterinary science, but not animal science, is often excluded from the definition.)

List of Countries by Agricultural Output

Largest countries by agricultural output according to IMF and CIA World Factbook, 2015	
Economy	**Countries by agricultural output in 2015 (billions in USD)**
(01) China	1,088
(02) India	413
(—) *European Union*	333
(03) United States	290
(04) Indonesia	127
(05) Brazil	110
(06) Nigeria	106
(07) Pakistan	63
(08) Turkey	62
(09) Argentina	59
(10) Japan	51
(11) Egypt	47
(12) Thailand	47
(13) Russia	47
(14) Australia	46
(15) Mexico	43
(16) France	42
(17) Italy	41
(18) Spain	39
(19) Vietnam	37
(20) Iran	36
The twenty largest countries by agricultural output in 2015, according to the IMF and CIA World Factbook.	

Energy and Agriculture

Since the 1940s, agricultural productivity has increased dramatically, due largely to the increased use of energy-intensive mechanization, fertilizers and pesticides. The vast majority of this energy input comes from fossil fuel sources. Between the 1960–65 measuring cycle and the cycle from 1986 to 1990, the Green Revolution transformed agriculture around the globe, with world grain production increasing significantly (between 70% and 390% for wheat and 60% to 150% for rice, depending on geographic area) as world population doubled. Modern agriculture's heavy reliance on petrochemicals and mechanization has raised concerns that oil shortages could increase costs and reduce agricultural output, causing food shortages.

Agriculture and food system share (%) of total energy consumption by three industrialized nations			
Country	Year	Agriculture (direct & indirect)	Food system
United Kingdom	2005	1.9	11
United States	1996	2.1	10
United States	2002	2.0	14
Sweden	2000	2.5	13

Modern or industrialized agriculture is dependent on fossil fuels in two fundamental ways: 1. direct consumption on the farm and 2. indirect consumption to manufacture inputs used on the farm. Direct consumption includes the use of lubricants and fuels to operate farm vehicles and machinery; and use of gasoline, liquid propane, and electricity to power dryers, pumps, lights, heaters, and coolers. American farms directly consumed about 1.2 exajoules (1.1 quadrillion BTU) in 2002, or just over 1% of the nation's total energy.

Indirect consumption is mainly oil and natural gas used to manufacture fertilizers and pesticides, which accounted for 0.6 exajoules (0.6 quadrillion BTU) in 2002. The natural gas and coal consumed by the production of nitrogen fertilizer can account for over half of the agricultural energy usage. China utilizes mostly coal in the production of nitrogen fertilizer, while most of Europe uses large amounts of natural gas and small amounts of coal. According to a 2010 report published by The Royal Society, agriculture is increasingly dependent on the direct and indirect input of fossil fuels. Overall, the fuels used in agriculture vary based on several factors, including crop, production system and location. The energy used to manufacture farm machinery is also a form of indirect agricultural energy consumption. Together, direct and indirect consumption by US farms accounts for about 2% of the nation's energy use. Direct and indirect energy consumption by U.S. farms peaked in 1979, and has gradually declined over the past 30 years. Food systems encompass not just agricultural production, but also off-farm processing, packaging, transporting, marketing, consumption, and disposal of food and food-related items. Agriculture accounts for less than one-fifth of food system energy use in the US.

Mitigation of Effects of Petroleum Shortages

In the event of a petroleum shortage, organic agriculture can be more attractive than conventional practices that use petroleum-based pesticides, herbicides, or fertilizers. Some studies using modern organic-farming methods have reported yields equal to or higher than those available from conventional farming. In the aftermath of the fall of the Soviet Union, with shortages of conventional petroleum-based inputs, Cuba made use of mostly organic practices, including biopesticides, plant-based pesticides and sustainable cropping practices, to feed its populace. However, organic farming may be more labor-intensive and would require a shift of the workforce from urban to rural areas. The reconditioning of soil to restore organic matter lost during the use of monoculture agriculture techniques is important to provide a reservoir of plant-available nutrients, to maintain texture, and to minimize erosion.

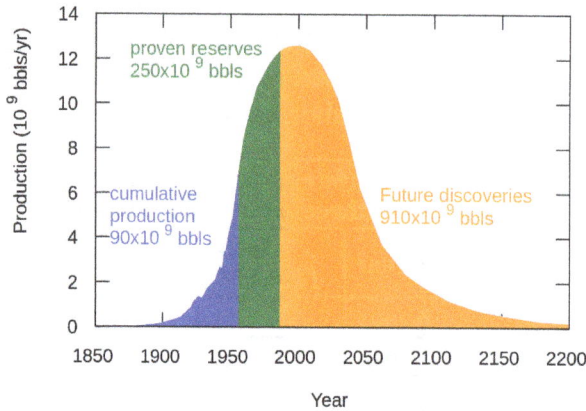

M. King Hubbert's prediction of world petroleum production rates.
Modern agriculture is totally reliant on petroleum energy

It has been suggested that rural communities might obtain fuel from the biochar and synfuel process, which uses agricultural *waste* to provide charcoal fertilizer, some fuel *and* food, instead of the normal food vs. fuel debate. As the synfuel would be used on-site, the process would be more efficient and might just provide enough fuel for a new organic-agriculture fusion.

It has been suggested that some transgenic plants may some day be developed which would allow for maintaining or increasing yields while requiring fewer fossil-fuel-derived inputs than conventional crops. The possibility of success of these programs is questioned by ecologists and economists concerned with unsustainable GMO practices such as terminator seeds. While there has been some research on sustainability using GMO crops, at least one prominent multi-year attempt by Monsanto Company has been unsuccessful, though during the same period traditional breeding techniques yielded a more sustainable variety of the same crop.

Policy

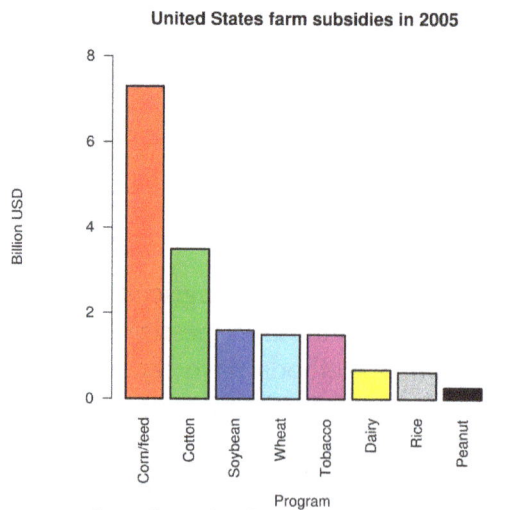

From a Congressional Budget Office report

Agricultural policy is the set of government decisions and actions relating to domestic agriculture and imports of foreign agricultural products. Governments usually implement agricultural policies with the goal of achieving a specific outcome in the domestic agricultural product markets. Some overarching themes include risk management and adjustment (including policies related to climate change, food safety and natural disasters), economic stability (including policies related to taxes), natural resources and environmental sustainability (especially water policy), research and development, and market access for domestic commodities (including relations with global organizations and agreements with other countries). Agricultural policy can also touch on food quality, ensuring that the food supply is of a consistent and known quality, food security, ensuring that the food supply meets the population's needs, and conservation. Policy programs can range from financial programs, such as subsidies, to encouraging producers to enroll in voluntary quality assurance programs.

There are many influences on the creation of agricultural policy, including consumers, agribusiness, trade lobbies and other groups. Agribusiness interests hold a large amount of influence over policy making, in the form of lobbying and campaign contributions. Political action groups, including those interested in environmental issues and labor unions, also provide influence, as do lobbying organizations representing individual agricultural commodities. The Food and Agriculture Organization of the United Nations (FAO) leads international efforts to defeat hunger and provides a forum for the negotiation of global agricultural regulations and agreements. Dr. Samuel Jutzi, director of FAO's animal production and health division, states that lobbying by large corporations has stopped reforms that would improve human health and the environment. For example, proposals in 2010 for a voluntary code of conduct for the livestock industry that would have provided incentives for improving standards for health, and environmental regulations, such as the number of animals an area of land can support without long-term damage, were successfully defeated due to large food company pressure.

Agricultural Science

Agricultural science is a broad multidisciplinary field of biology that encompasses the parts of exact, natural, economic and social sciences that are used in the practice and understanding of agriculture. (Veterinary science, but not animal science, is often excluded from the definition.)

Agriculture, Agricultural Science, and Agronomy

The three terms are often confused. However, they cover different concepts:

- Agriculture is the set of activities that transform the environment for the production of animals and plants for human use. Agriculture concerns techniques, including the application of agronomic research.

- Agronomy is research and development related to studying and improving plant-based crops.

Agricultural sciences include research and development on:

- Production techniques (e.g., irrigation management, recommended nitrogen inputs)

- Improving agricultural productivity in terms of quantity and quality (e.g., selection of drought-resistant crops and animals, development of new pesticides, yield-sensing technologies, simulation models of crop growth, in-vitro cell culture techniques)

- Minimizing the effects of pests (weeds, insects, pathogens, nematodes) on crop or animal production systems.

- Transformation of primary products into end-consumer products (e.g., production, preservation, and packaging of dairy products)

- Prevention and correction of adverse environmental effects (e.g., soil degradation, waste management, bioremediation)

- Theoretical production ecology, relating to crop production modeling

- Traditional agricultural systems, sometimes termed subsistence agriculture, which feed most of the poorest people in the world. These systems are of interest as they sometimes retain a level of integration with natural ecological systems greater than that of industrial agriculture, which may be more sustainable than some modern agricultural systems.

- Food production and demand on a global basis, with special attention paid to the major producers, such as China, India, Brazil, the USA and the EU.

- Various sciences relating to agricultural resources and the environment (e.g. soil science, agroclimatology); biology of agricultural crops and animals (e.g. crop science, animal science and their included sciences, e.g. ruminant nutrition, farm animal welfare); such fields as agricultural economics and rural sociology; various disciplines encompassed in agricultural engineering.

Agricultural Biotechnology

Agricultural biotechnology is a specific area of agricultural science involving the use of scientific tools and techniques, including genetic engineering, molecular markers, molecular diagnostics, vaccines, and tissue culture, to modify living organisms: plants, animals, and microorganisms.

Fertilizer

One of the most common yield reducers is because of fertilizer not being applied in slightly higher quantities during transition period, the time it takes the soil to rebuild its aggregates and organic matter. Yields will decrease temporarily because of nitrogen being immobilized in the crop residue, which can take a few months to several years to decompose, depending on the crop's C to N ratio and the local environment

A Local Science

With the exception of theoretical agronomy, research in agronomy, more than in any other field, is strongly related to local areas. It can be considered a science of ecoregions, because it is closely linked to soil properties and climate, which are never exactly the same from one place to another. Many people think an agricultural production system relying on local weather, soil characteristics, and specific crops has to be studied locally. Others feel a need to know and understand production systems in as many areas as possible, and the human dimension of interaction with nature.

History of Agricultural Science

Agricultural science began with Gregor Mendel's genetic work, but in modern terms might be better dated from the chemical fertilizer outputs of plant physiological understanding in 18th-century Germany. In the United States, a scientific revolution in agriculture began with the Hatch Act of 1887, which used the term "agricultural science". The Hatch Act was driven by farmers' interest in knowing the constituents of early artificial fertilizer. The Smith-Hughes Act of 1917 shifted agricultural education back to its vocational roots, but the scientific foundation had been built. After 1906, public expenditures on agricultural research in the US exceeded private expenditures for the next 44 years.

Intensification of agriculture since the 1960s in developed and developing countries, often referred to as the Green Revolution, was closely tied to progress made in selecting and improving crops and animals for high productivity, as well as to developing additional inputs such as artificial fertilizers and phytosanitary products.

As the oldest and largest human intervention in nature, the environmental impact of agriculture in general and more recently intensive agriculture, industrial development, and population growth have raised many questions among agricultural scientists and have led to the development and emergence of new fields. These include technological fields that assume the solution to technological problems lies in better technology, such as integrated pest management, waste treatment technologies, landscape architecture, genomics, and agricultural philosophy fields that include references to food production as something essentially different from non-essential economic 'goods'. In fact, the interaction between these two approaches provide a fertile field for deeper understanding in agricultural science.

New technologies, such as biotechnology and computer science (for data processing and storage), and technological advances have made it possible to develop new research fields, including genetic engineering, agrophysics, improved statistical analysis, and precision farming. Balancing these, as above, are the natural and human sciences of agricultural science that seek to understand the human-nature interactions of traditional agriculture, including interaction of religion and agriculture, and the non-material components of agricultural production systems.

Prominent Agricultural Scientists

Norman Borlaug, father of the Green Revolution.

- Robert Bakewell
- Norman Borlaug
- Luther Burbank
- George Washington Carver
- René Dumont
- Sir Albert Howard
- Kailas Nath Kaul
- Justus von Liebig
- Jay Lush
- Gregor Mendel
- Louis Pasteur
- M. S. Swaminathan
- Jethro Tull
- Artturi Ilmari Virtanen
- Eli Whitney
- Sewall Wright

Agricultural Science and Agriculture Crisis

Agriculture sciences seek to feed the world's population while preventing biosafety problems that may affect human health and the environment. This requires promoting good management of natural resources and respect for the environment, and increasingly concern for the psychological wellbeing of all concerned in the food production and consumption system.

Economic, environmental, and social aspects of agriculture sciences are subjects of ongoing debate. Recent crises (such as avian influenza, mad cow disease and issues such as the use of genetically modified organisms) illustrate the complexity and importance of this debate.

Fields or Related Disciplines

• Agricultural biotechnology	• Aquaculture
• Agricultural chemistry	• Biological engineering
• Agricultural diversification	o Genetic engineering
• Agricultural education	• Nematology
• Agricultural economics	• Microbiology
• Agricultural engineering	o Plant pathology
• Agricultural geography	o Range Management

- Agricultural philosophy
- Agricultural marketing
- Agricultural soil science
- Agrophysics
- Animal science
  - o Animal breeding
  - o Animal husbandry
  - o Animal nutrition
- Agronomy
  - o Botany
  - o Theoretical production ecology
  - o Horticulture
  - o Plant breeding
  - o Plant fertilization

- Environmental science
- Entomology
- Food science
  - o Human nutrition
- Irrigation and water management
- Soil science
  - o Agrology
- Waste management
- Weed science

Food Industry

Packaged food aisles at an American grocery store

Parmigiano reggiano Cheese in a modern factory

Hens in a battery cage in Brazil, an example of intensive animal farming

The food industry is a complex, global collective of diverse businesses that supply most of the food consumed by the world population. Only subsistence farmers, those who survive on what they grow, can be considered outside of the scope of the modern food industry.

The food industry includes:

- Agriculture: raising of crops and livestock, and seafood

- Manufacturing: agrichemicals, agricultural construction, farm machinery and supplies, seed, etc.

- Food processing: preparation of fresh products for market, and manufacture of prepared food products

- Marketing: promotion of generic products (e.g., milk board), new products, advertising, marketing campaigns, packaging, public relations, etc.

- Wholesale and distribution: logistics, transportation, warehousing

- Foodservice (which includes Catering)

- Grocery, farmers' markets, public markets and other retailing

- Regulation: local, regional, national, and international rules and regulations for food production and sale, including food quality, food safety, marketing/advertising, and industry lobbying activities

- Education: academic, consultancy, vocational

- Research and development: food technology

- Financial services: credit, insurance

Definitions

It is challenging to find an inclusive way to cover all aspects of food production and sale. The Food Standards Agency, a government body in India, describes it thusly:

> "...the whole food industry – from farming and food production, packaging and distribution, to retail and catering."

The Economic Research Service of the USDA uses the term *food system* to describe the same thing:

> "The U.S. food system is a complex network of farmers and the industries that link to them. Those links include makers of farm equipment and chemicals as well as firms that provide services to agribusinesses, such as providers of transportation and financial services. The system also includes the food marketing industries that link farms to consumers, and which include food and fiber processors, wholesalers, retailers, and foodservice establishments."

Agriculture and Agronomy

Agriculture is the process of producing food, feeding products, fiber and other desired products by the cultivation of certain plants and the raising of domesticated animals (livestock). The practice of agriculture is also known as "farming". Scientists, inventors, and others devoted to improving farming methods and implements are also said to be engaged in agriculture. 1 in 3 people worldwide are employed in agriculture, yet it only contributes 3% to global GDP.

A soybean field in Junin, Argentina

Agronomy is the science and technology of producing and using plants for food, fuel, fibre, and land reclamation. Agronomy encompasses work in the areas of plant genetics, plant physiology, meteorology, and soil science. Agronomy is the application of a combination of sciences. Agronomists today are involved with many issues including producing food, creating healthier food, managing environmental impact of agriculture, and extracting energy from plants.

Food Processing

Food processing includes the methods and techniques used to transform raw ingredients into food for human consumption. Food processing takes clean, harvested or slaughtered and butchered components and uses them to produce marketable food products. There are several different ways in which food can be produced.

Packaged meat in a supermarket

One off production: This method is used when customers make an order for something to be made to their own specifications, for example a wedding cake. The making of one-off products could take days depending on how intricate the design is.

Batch production: This method is used when the size of the market for a product is not clear, and where there is a range within a product line. A certain number of the same goods will be produced to make up a batch or run, for example a bakery may bake a limited number of cupcakes. This method involves estimating consumer demand.

Mass production: This method is used when there is a mass market for a large number of identical products, for example chocolate bars, ready meals and canned food. The product passes from one stage of production to another along a production line.

Just-in-time (JIT) (production): This method of production is mainly used in restaurants. All components of the product are available in-house and the customer chooses what they want in the product. It is then prepared in a kitchen, or in front of the buyer as in sandwich delicatessens, pizzerias, and sushi bars.

Wholesale and Distribution

A vast global cargo network connects the numerous parts of the industry. These include suppliers, manufacturers, warehousers, retailers and the end consumers. Wholesale markets for fresh food products have tended to decline in importance in urbanizing countries, including Latin America and some Asian countries as a result of the growth of supermarkets, which procure directly from farmers or through preferred suppliers, rather than going through markets.

The constant and uninterrupted flow of product from distribution centers to store locations is a critical link in food industry operations. Distribution centers run more efficiently, throughput can be increased, costs can be lowered, and manpower better utilized if the proper steps are taken when setting up a material handling system in a warehouse.

A foodservice truck at a loading dock. Trucks commonly distribute food products to commercial businesses and organizations.

Retail

With worldwide urbanization, food buying is increasingly removed from food production. During the 20th century, the supermarket became the defining retail element of the food industry. There, tens of thousands of products are gathered in one location, in continuous, year-round supply.

Food preparation is another area where the change in recent decades has been dramatic. Today, two food industry sectors are in apparent competition for the retail food dollar. The grocery industry sells fresh and largely raw products for consumers to use as ingredients in home cooking. The food service industry by contrast offers prepared food, either as finished products, or as partially prepared components for final "assembly". Restaurants, cafes, bakeries and mobile food trucks provide opportunities for consumers to purchase food.

Food Industry Technologies

The Passaic Agricultural Chemical Works, an agrochemical company, in Newark, New Jersey, 1876

Modern food production is defined by sophisticated technologies. These include many areas. Agricultural machinery, originally led by the tractor, has practically eliminated human labor in many areas of production. Biotechnology is driving much change, in areas as diverse as agrochemicals, plant breeding and food processing. Many other types of technology are also involved, to the point where it is hard to find an area that does not have a direct impact on the food industry. As in other fields, computer technology is also a central force, with computer networks and specialized software providing the support infrastructure to allow global movement of the myriad components involved.

Marketing

As consumers grow increasingly removed from food production, the role of product creation, advertising, and publicity become the primary vehicles for information about food. With processed food as the dominant category, marketers have almost infinite possibilities in product creation.

Labor and Education

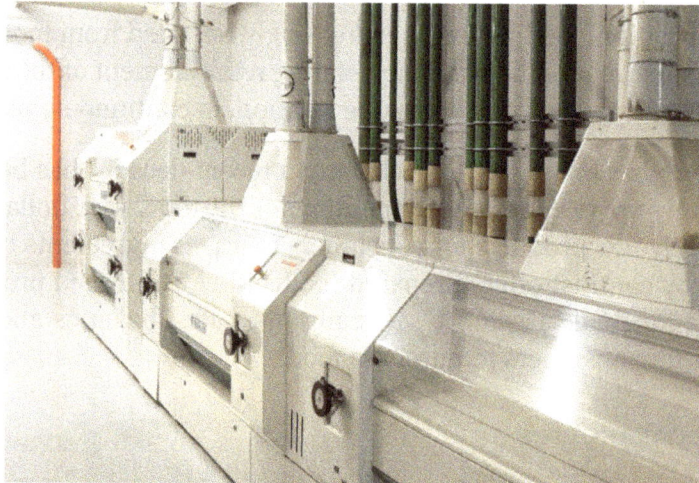

Some equipment at Tartu Mill, the largest grain milling company in the Baltic states. Modern food processing factories are often highly automated and need few workers.

Until the last 100 years, agriculture was labor intensive. Farming was a common occupation and millions of people were involved in food production. Farmers, largely trained from generation to generation, carried on the family business. That situation has changed dramatically today. In America in 1870, 70-80 percent of the US population was employed in agriculture. As of 2008, less than 2 percent of the population is directly employed in agriculture., and about 80% of the population lives in cities. The food industry as a complex whole requires an incredibly wide range of skills. Several hundred occupation types exist within the food industry.

By Country

- Food industry in Bangladesh

- Ministry of Agriculture and Food Industry (Moldova)

- Food industry of South Africa

Food Processing

Industrial cheese production

Michael Foods egg-processing plant in Wakefield, Nebraska

Food processing is the transformation of raw ingredients, by physical or chemical means into food, or of food into other forms. Food processing combines raw food ingredients to produce marketable food products that can be easily prepared and served by the consumer. Food processing typically involves activities such as mincing and macerating, liquefaction, emulsification, and cooking (such as boiling, broiling, frying, or grilling); pickling, pasteurization, and many other kinds of preservation; and canning or other packaging. (Primary-processing such as dicing or slicing, freezing or drying when leading to secondary products are also included.)

History

Food processing dates back to the prehistoric ages when crude processing incorporated fermenting, sun drying, preserving with salt, and various types of cooking (such as roasting, smoking, steaming, and oven baking), Such basic food processing involved chemical enzymatic changes to the basic structure of food in its natural form, as well served to build a barrier against surface microbial activity that caused rapid decay. Salt-preservation was especially common for foods that constituted warrior and sailors' diets until the introduction of canning methods. Evidence for the existence of these methods can be found in the writings of the ancient Greek, Chaldean, Egyptian and Roman civilizations as well as archaeological evidence from Europe, North and South America and Asia. These tried and tested processing techniques remained essentially the same until the advent of the industrial revolution. Examples of ready-meals also date back to before the preindustrial revolution, and include dishes such as Cornish pasty and Haggis. Both during ancient times and today in modern society these are considered processed foods.

Grain silos in Ardrossan, Scotland

Modern food processing technology developed in the 19th and 20th centuries was developed in a large part to serve military needs. In 1809 Nicolas Appert invented a hermetic bottling technique that would preserve food for French troops which ultimately contributed to the development of tinning, and subsequently canning by Peter Durand in 1810. Although initially expensive and somewhat hazardous due to the lead used in cans, canned goods would later become a staple around the world. Pasteurization, discovered by Louis Pasteur in 1864, improved the quality of preserved foods and introduced the wine, beer, and milk preservation.

A form of pre-made split-pea soup that has become traditional

In the 20th century, World War II, the space race and the rising consumer society in developed countries (including the United States) contributed to the growth of food processing with such advances as spray drying, juice concentrates, freeze drying and the introduction of artificial sweeteners, colouring agents, and such preservatives as sodium benzoate. In the late 20th century, products such as dried instant soups, reconstituted fruits and juices, and self cooking meals such as MRE food ration were developed.

In western Europe and North America, the second half of the 20th century witnessed a rise in the pursuit of convenience. Food processing companies marketed their products especially towards middle-class working wives and mothers. Frozen foods (often credited to Clarence Birdseye) found their success in sales of juice concentrates and "TV dinners". Processors utilised the perceived value of time to appeal to the postwar population, and this same appeal contributes to the success of convenience foods today.

Benefits and Drawbacks

Benefits

Benefits of food processing include toxin removal, preservation, easing marketing and distribution tasks, and increasing food consistency. In addition, it increases yearly availability of many foods, enables transportation of delicate perishable foods across long distances and makes many kinds of foods safe to eat by de-activating spoilage and pathogenic micro-organisms. Modern supermarkets would not exist without modern food processing techniques, and long voyages would not be possible.

Processed seafood - fish, squid, prawn balls and simulated crab sticks (surimi)

Processed foods are usually less susceptible to early spoilage than fresh foods and are better suited for long distance transportation from the source to the consumer. When they were first introduced, some processed foods helped to alleviate food shortages and improved the overall nutrition of populations as it made many new foods available to the masses.

Processing can also reduce the incidence of food borne disease. Fresh materials, such as fresh produce and raw meats, are more likely to harbour pathogenic micro-organisms (e.g. Salmonella) capable of causing serious illnesses.

The extremely varied modern diet is only truly possible on a wide scale because of food processing. Transportation of more exotic foods, as well as the elimination of much hard labour gives the modern eater easy access to a wide variety of food unimaginable to their ancestors.

The act of processing can often improve the taste of food significantly.

Mass production of food is much cheaper overall than individual production of meals from raw ingredients. Therefore, a large profit potential exists for the manufacturers and suppliers of processed food products. Individuals may see a benefit in convenience, but rarely see any direct financial cost benefit in using processed food as compared to home preparation.

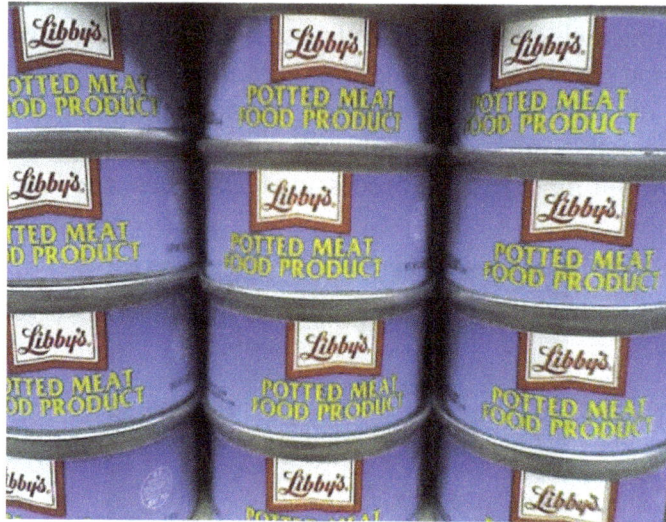

Libby's brand 'Potted Meat Food Product'

Processed food freed people from the large amount of time involved in preparing and cooking "natural" unprocessed foods. The increase in free time allows people much more choice in life style than previously allowed. In many families the adults are working away from home and therefore there is little time for the preparation of food based on fresh ingredients. The food industry offers products that fulfill many different needs: e.g. fully prepared ready meals that can be heated up in the microwave oven within a few minutes.

Modern food processing also improves the quality of life for people with allergies, diabetics, and other people who cannot consume some common food elements. Food processing can also add extra nutrients such as vitamins.

Drawbacks

Any processing of food can affect its nutritional density. The amount of nutrients lost depends on the food and processing method. For example, heat destroys vitamin C. Therefore, canned fruits possess less vitamin C than their fresh alternatives. The USDA conducted a study in 2004, creating a nutrient retention table for several foods.

New research highlighting the importance to human health of a rich microbial environment in the intestine indicates that abundant food processing (not fermentation of foods) endangers that environment.

Meat packages in a Roman supermarket

Using food additives represents another safety concern. The health risks of any given additive vary greatly from person to person; for example using sugar as an additive endangers diabetics. In the European Union, only European Food Safety Authority (EFSA) approved food additives (e.g., sweeteners, preservatives, stabilizers) are permitted at specified levels for use in food products. Approved additives receive an E number (E for Europe), simplifying communication about food additives included in the ingredients' list for all the different languages spoken in the EU. Certain additives can also result in an addiction to a particular food item. As effects of chemical additives are learnt, changes to laws and regulatory practices are made to make such processed foods more safe.

Food processing is typically a mechanical process that utilizes large mixing, grinding, chopping and emulsifying equipment in the production process. These processes inherently introduce a number of contamination risks. As a mixing bowl or grinder is used over time the food contact parts will tend to fail and fracture. This type of failure will introduce into the product stream small to large metal contaminants. Further processing of these metal fragments will result in downstream equipment failure and the risk of ingestion by the consumer. Food manufacturers utilize industrial metal detectors to detect and reject automatically any metal fragment. Large food processors will utilize many metal detectors within the processing stream to reduce both damage to processing machinery as well as risk to consumer health.

Typical Maximum Nutrient Losses due to cooking					
Vitamin & Minerals	**Freeze**	**Dry**	**Cook**	**Cook+Drain**	**Reheat**
Vitamin A	5%	50%	25%	35%	10%
Vit A- Retinol Activity Equivalent	5%	50%	25%	35%	10%
Vit A- Alpha Carotene	5%	50%	25%	35%	10%
Vit A- Beta Carotene	5%	50%	25%	35%	10%
Vit A- Beta Cryptoxanthin	5%	50%	25%	35%	10%
Vit A- Lycopene	5%	50%	25%	35%	10%
Vit A- Lutein+Zeaxanthin	5%	50%	25%	35%	10%
Vitamin C	30%	80%	50%	75%	50%

Thiamin	5%	30%	55%	70%	40%
Riboflavin	0%	10%	25%	45%	5%
Niacin	0%	10%	40%	55%	5%
Vitamin B6	0%	10%	50%	65%	45%
Folate	5%	50%	70%	75%	30%
Food Folate	5%	50%	70%	75%	30%
Folic Acid	5%	50%	70%	75%	30%
Vitamin B12	0%	0%	45%	50%	45%
Calcium	5%	0%	20%	25%	0%
Iron	0%	0%	35%	40%	0%
Magnesium	0%	0%	25%	40%	0%
Phosphorus	0%	0%	25%	35%	0%
Potassium	10%	0%	30%	70%	0%
Sodium	0%	0%	25%	55%	0%
Zinc	0%	0%	25%	25%	0%
Copper	10%	0%	40%	45%	0%

Performance Parameters for Food Processing

Factory automation - robotics palettizing bread

When designing processes for the food industry the following performance parameters may be taken into account:

- Hygiene, e.g. measured by number of micro-organisms per mL of finished product

- Energy efficiency measured e.g. by "ton of steam per ton of sugar produced"

- Minimization of waste, measured e.g. by "percentage of peeling loss during the peeling of potatoes"

- Labour used, measured e.g. by "number of working hours per ton of finished product"

- Minimization of cleaning stops measured e.g. by "number of hours between cleaning stops"

De-agglomerating Batter Mixes in Food Processing

Problems often occur during preparation of batter mixes because flour and other powdered ingredients tend to form lumps or agglomerates as they are being mixed during production. A conventional mixer/agitator cannot break down these agglomerates, resulting in a lumpy batter. If lumpy batter is used to enrobe products, it causes an unsatisfactory appearance with misshapen or oversize products that do not fit properly into packaging. This can force production to a standstill. Furthermore, batter mix is generally recirculated from an enrobing system back to a holding vessel; lumps then have a tendency to build up, reducing the flow of material and raising potential sanitation issues.

Using a high shear in-line mixer in place of a conventional agitator or mixer can quickly solve problems of agglomeration with dry ingredients. A single pass through a self-pumping, in-line mixer adds high shear to batter, which de-agglomerates the mix, resulting in a homogeneous, smooth batter. With a consistent, smooth batter, finished product appearance is improved; the effectiveness and hygiene of the recirculation system is increased; and a better yield of raw materials is achieved. By increasing overall product quality, the amount of raw materials needed is decreased, thereby lowering manufacturing costs and increasing shelf life. Increased shelf life is achieved by creating and maintaining an emulsion, often by adding a food stabilizer.

High shear in-line mixers process food to be made faster and cheaper while increasing consistency of the finished food. Powder and liquid mixing systems are capable of rapidly incorporating large quantities of powders at high concentrations – agglomerate free and fully hydrated. Advances in technology have made processing equipment easy to clean, leading to a much safer processed food.

Trends in Modern Food Processing

Women working in a cannery

Dried bananas packaged in Ban Bang Krathum, Bang Krathum, Phitsanulok, Thailand

Health

- Reduction of fat content in final product by using baking instead of deep-frying in the production of potato chips, another processed food.

- Maintaining the natural taste of the product by using less artificial sweetener than was used before.

Hygiene

The rigorous application of industry and government endorsed standards to minimise possible risk and hazards. The international standard adopted is HACCP.

Efficiency

- Rising energy costs lead to increasing usage of energy-saving technologies, e.g. frequency converters on electrical drives, heat insulation of factory buildings and heated vessels, energy recovery systems, keeping a single fish frozen all the way from China to Switzerland.

- Factory automation systems (often Distributed control systems) reduce personnel costs and may lead to more stable production results.

Industries

Food processing industries and practices include the following:

- Cannery
- Fish processing
- Food packaging plant
- Industrial rendering
- Meat packing plant
- Slaughterhouse
- Sugar industry

Medicine

Medicine is the science and practice of the diagnosis, treatment, and prevention of disease. The word *medicine* is derived from Latin *medicus*, meaning "a physician". Medicine encompasses a variety of health care practices evolved to maintain and restore health by the prevention and treatment of illness. Contemporary medicine applies biomedical sciences, biomedical research, genetics, and medical technology to diagnose, treat, and prevent injury and disease, typically through pharmaceuticals or surgery, but also through therapies as diverse as psychotherapy, external splints and traction, medical devices, biologics, and ionizing radiation, amongst others.

Medicine has existed for thousands of years, during most of which it was an art (an area of skill and knowledge) frequently having connections to the religious and philosophical beliefs of local culture. For example, a medicine man would apply herbs and say prayers for healing, or an ancient philosopher and physician would apply bloodletting according to the theories of humorism. In recent centuries, since the advent of modern science, most medicine has become a combination of art and science (both basic and applied, under the umbrella of medical science). While stitching technique for sutures is an art learned through practice, the knowledge of what happens at the cellular and molecular level in the tissues being stitched arises through science.

Prescientific forms of medicine are now known as traditional medicine and folk medicine. They remain commonly used with or instead of scientific medicine and are thus called alternative medicine. For example, evidence on the effectiveness of acupuncture is "variable and inconsistent" for any condition, but is generally safe when done by an appropriately trained practitioner. In contrast, treatments outside the bounds of safety and efficacy are termed quackery.

Clinical Practice

The Doctor, by Sir Luke Fildes (1891)

Medical availability and clinical practice varies across the world due to regional differences in culture and technology. Modern scientific medicine is highly developed in the Western world, while in developing countries such as parts of Africa or Asia, the population may rely more heavily on traditional medicine with limited evidence and efficacy and no required formal training for practitioners. Even in the developed world however, evidence-based medicine is not universally used in clinical practice; for example, a 2007 survey of literature reviews found that about 49% of the interventions lacked sufficient evidence to support either benefit or harm.

In modern clinical practice, doctors personally assess patients in order to diagnose, treat, and prevent disease using clinical judgment. The doctor-patient relationship typically begins an interaction with an examination of the patient's medical history and medical record, followed by a medical interview and a physical examination. Basic diagnostic medical devices (e.g. stethoscope, tongue depressor) are typically used. After examination for signs and interviewing for symptoms, the doctor may order medical tests (e.g. blood tests), take a biopsy, or prescribe pharmaceutical drugs or other therapies. Differential diagnosis methods help to rule out conditions based on the information provided. During the encounter, properly informing the patient of all relevant facts is an important part of the relationship and the development of trust. The medical encounter is then documented in the medical record, which is a legal document in many jurisdictions. Follow-ups may be shorter but follow the same general procedure, and specialists follow a similar process. The diagnosis and treatment may take only a few minutes or a few weeks depending upon the complexity of the issue.

The components of the medical interview and encounter are:

- Chief complaint (CC): the reason for the current medical visit. These are the 'symptoms.' They are in the patient's own words and are recorded along with the duration of each one. Also called 'chief concern' or 'presenting complaint'.

- History of present illness (HPI): the chronological order of events of symptoms and further clarification of each symptom. Distinguishable from history of previous illness, often called past medical history (PMH). Medical history comprises HPI and PMH.

- Current activity: occupation, hobbies, what the patient actually does.

- Medications (Rx): what drugs the patient takes including prescribed, over-the-counter, and home remedies, as well as alternative and herbal medicines/herbal remedies. Allergies are also recorded.

- Past medical history (PMH/PMHx): concurrent medical problems, past hospitalizations and operations, injuries, past infectious diseases and/or vaccinations, history of known allergies.

- Social history (SH): birthplace, residences, marital history, social and economic status, habits (including diet, medications, tobacco, alcohol).

- Family history (FH): listing of diseases in the family that may impact the patient. A family tree is sometimes used.

- Review of systems (ROS) or *systems inquiry*: a set of additional questions to ask, which may be missed on HPI: a general enquiry (have you noticed any weight loss, change in sleep quality, fevers, lumps and bumps? etc.), followed by questions on the body's main organ systems (heart, lungs, digestive tract, urinary tract, etc.).

The physical examination is the examination of the patient for medical signs of disease, which are objective and observable, in contrast to symptoms which are volunteered by the patient and not necessarily objectively observable. The healthcare provider uses the senses of sight, hearing, touch, and sometimes smell (e.g., in infection, uremia, diabetic ketoacidosis). Four actions are the basis of physical examination: inspection, palpation (feel), percussion (tap to determine resonance characteristics), and auscultation (listen), generally in that order although auscultation occurs prior to percussion and palpation for abdominal assessments.

The clinical examination involves the study of:

- Vital signs including height, weight, body temperature, blood pressure, pulse, respiration rate, and hemoglobin oxygen saturation

- General appearance of the patient and specific indicators of disease (nutritional status, presence of jaundice, pallor or clubbing)

- Skin

- Head, eye, ear, nose, and throat (HEENT)

- Cardiovascular (heart and blood vessels)

- Respiratory (large airways and lungs)

- Abdomen and rectum

- Genitalia (and pregnancy if the patient is or could be pregnant)

- Musculoskeletal (including spine and extremities)

- Neurological (consciousness, awareness, brain, vision, cranial nerves, spinal cord and peripheral nerves)

- Psychiatric (orientation, mental state, evidence of abnormal perception or thought).

It is to likely focus on areas of interest highlighted in the medical history and may not include everything listed above.

The treatment plan may include ordering additional medical laboratory tests and medical imaging studies, starting therapy, referral to a specialist, or watchful observation. Follow-up may be advised. Depending upon the health insurance plan and the managed care system, various forms of "utilization review", such as prior authorization of tests, may place barriers on accessing expensive services.

The medical decision-making (MDM) process involves analysis and synthesis of all the above data to come up with a list of possible diagnoses (the differential diagnoses), along with an idea of what needs to be done to obtain a definitive diagnosis that would explain the patient's problem.

On subsequent visits, the process may be repeated in an abbreviated manner to obtain any new history, symptoms, physical findings, and lab or imaging results or specialist consultations.

Institutions

Contemporary medicine is in general conducted within health care systems. Legal, credentialing and financing frameworks are established by individual governments, augmented on occasion by international organizations, such as churches. The characteristics of any given health care system have significant impact on the way medical care is provided.

From ancient times, Christian emphasis on practical charity gave rise to the development of systematic nursing and hospitals and the Catholic Church today remains the largest non-government provider of medical services in the world. Advanced industrial countries (with the exception of the United States) and many developing countries provide medical services through a system of universal health care that aims to guarantee care for all through a single-payer health care system, or compulsory private or co-operative health insurance. This is intended to ensure that the entire population has access to medical care on the basis of need rather than ability to pay. Delivery may be via private medical practices or by state-owned hospitals and clinics, or by charities, most commonly by a combination of all three.

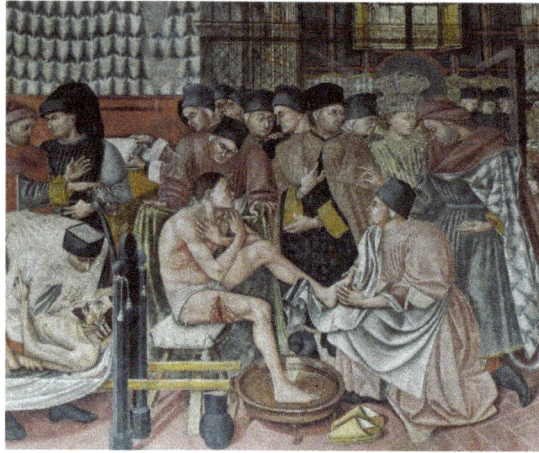

The Hospital of Santa Maria della Scala, fresco by Domenico di Bartolo, 1441–1442

Most tribal societies provide no guarantee of healthcare for the population as a whole. In such societies, healthcare is available to those that can afford to pay for it or have self-insured it (either directly or as part of an employment contract) or who may be covered by care financed by the government or tribe directly.

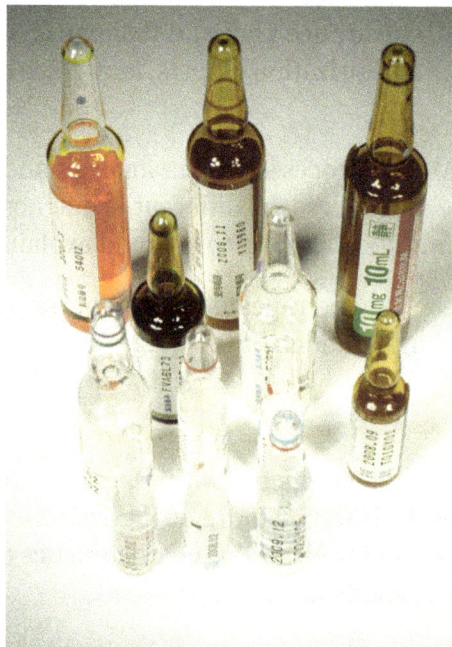

Modern drug ampoules

Transparency of information is another factor defining a delivery system. Access to information on conditions, treatments, quality, and pricing greatly affects the choice by patients/consumers and, therefore, the incentives of medical professionals. While the US healthcare system has come under fire for lack of openness, new legislation may encourage greater openness. There is a perceived tension between the need for transparency on the one hand and such issues as patient confidentiality and the possible exploitation of information for commercial gain on the other.

Delivery

Provision of medical care is classified into primary, secondary, and tertiary care categories.

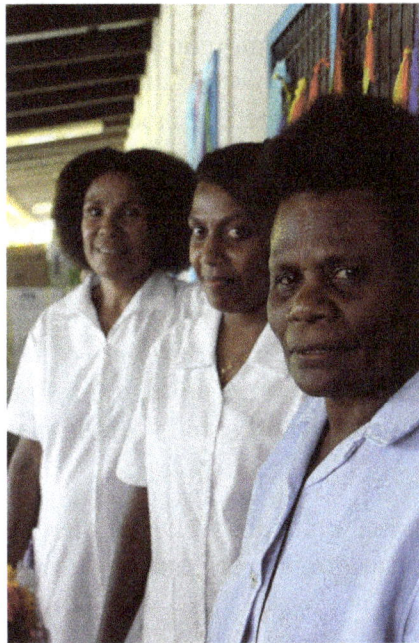

Nurses in Kokopo, East New Britain, Papua New Guinea

Primary care medical services are provided by physicians, physician assistants, nurse practitioners, or other health professionals who have first contact with a patient seeking medical treatment or care. These occur in physician offices, clinics, nursing homes, schools, home visits, and other places close to patients. About 90% of medical visits can be treated by the primary care provider. These include treatment of acute and chronic illnesses, preventive care and health education for all ages and both sexes.

Secondary care medical services are provided by medical specialists in their offices or clinics or at local community hospitals for a patient referred by a primary care provider who first diagnosed or treated the patient. Referrals are made for those patients who required the expertise or procedures performed by specialists. These include both ambulatory care and inpatient services, emergency rooms, intensive care medicine, surgery services, physical therapy, labor and delivery, endoscopy units, diagnostic laboratory and medical imaging services, hospice centers, etc. Some primary care providers may also take care of hospitalized patients and deliver babies in a secondary care setting.

Tertiary care medical services are provided by specialist hospitals or regional centers equipped with diagnostic and treatment facilities not generally available at local hospitals. These include trauma centers, burn treatment centers, advanced neonatology unit services, organ transplants, high-risk pregnancy, radiation oncology, etc.

Modern medical care also depends on information – still delivered in many health care settings on paper records, but increasingly nowadays by electronic means.

In low-income countries, modern healthcare is often too expensive for the average person. International healthcare policy researchers have advocated that "user fees" be removed in these areas to ensure access, although even after removal, significant costs and barriers remain.

Branches

Working together as an interdisciplinary team, many highly trained health professionals besides medical practitioners are involved in the delivery of modern health care. Examples include: nurses, emergency medical technicians and paramedics, laboratory scientists, pharmacists, podiatrists, physiotherapists, respiratory therapists, speech therapists, occupational therapists, radiographers, dietitians, and bioengineers, surgeons, surgeon's assistant, surgical technologist.

The scope and sciences underpinning human medicine overlap many other fields. Dentistry, while considered by some a separate discipline from medicine, is a medical field.

A patient admitted to the hospital is usually under the care of a specific team based on their main presenting problem, e.g., the Cardiology team, who then may interact with other specialties, e.g., surgical, radiology, to help diagnose or treat the main problem or any subsequent complications/developments.

Physicians have many specializations and subspecializations into certain branches of medicine, which are listed below. There are variations from country to country regarding which specialties certain subspecialties are in.

The main branches of medicine are:

- Basic sciences of medicine; this is what every physician is educated in, and some return to in biomedical research

- Medical specialties

- Interdisciplinary fields, where different medical specialties are mixed to function in certain occasions.

Basic Sciences

- *Anatomy* is the study of the physical structure of organisms. In contrast to *macroscopic* or *gross anatomy*, *cytology* and *histology* are concerned with microscopic structures.

- *Biochemistry* is the study of the chemistry taking place in living organisms, especially the structure and function of their chemical components.

- *Biomechanics* is the study of the structure and function of biological systems by means of the methods of Mechanics.

- *Biostatistics* is the application of statistics to biological fields in the broadest sense. A knowledge of biostatistics is essential in the planning, evaluation, and interpretation of medical research. It is also fundamental to epidemiology and evidence-based medicine.

- *Biophysics* is an interdisciplinary science that uses the methods of physics and physical chemistry to study biological systems.

- *Cytology* is the microscopic study of individual cells.

Louis Pasteur in his laboratory, 1885

- *Embryology* is the study of the early development of organisms.

- *Endocrinology* is the study of hormones and their effect throughout the body of animals.

- *Epidemiology* is the study of the demographics of disease processes, and includes, but is not limited to, the study of epidemics.

- *Genetics* is the study of genes, and their role in biological inheritance.

- *Histology* is the study of the structures of biological tissues by light microscopy, electron microscopy and immunohistochemistry.

- *Immunology* is the study of the immune system, which includes the innate and adaptive immune system in humans, for example.

- *Medical physics* is the study of the applications of physics principles in medicine.

- *Microbiology* is the study of microorganisms, including protozoa, bacteria, fungi, and viruses.

- *Molecular biology* is the study of molecular underpinnings of the process of replication, transcription and translation of the genetic material.

- *Neuroscience* includes those disciplines of science that are related to the study of the nervous system. A main focus of neuroscience is the biology and physiology of the human brain and spinal cord. Some related clinical specialties include neurology, neurosurgery and psychiatry.

- *Nutrition science* (theoretical focus) and *dietetics* (practical focus) is the study of the relationship of food and drink to health and disease, especially in determining an optimal diet. Medical nutrition therapy is done by dietitians and is prescribed for diabetes, cardiovascular diseases, weight and eating disorders, allergies, malnutrition, and neoplastic diseases.

- *Pathology as a science* is the study of disease—the causes, course, progression and resolution thereof.

- *Pharmacology* is the study of drugs and their actions.

- *Photobiology* is the study of the interactions between non-ionizing radiation and living organisms.

- *Physiology* is the study of the normal functioning of the body and the underlying regulatory mechanisms.

- *Radiobiology* is the study of the interactions between ionizing radiation and living organisms.

- *Toxicology* is the study of hazardous effects of drugs and poisons.

Specialties

In the broadest meaning of "medicine", there are many different specialties. In the UK, most specialities have their own body or college, which have its own entrance examination. These are collectively known as the Royal Colleges, although not all currently use the term "Royal". The development of a speciality is often driven by new technology (such as the development of effective anaesthetics) or ways of working (such as emergency departments); the new specialty leads to the formation of a unifying body of doctors and the prestige of administering their own examination.

Within medical circles, specialities usually fit into one of two broad categories: "Medicine" and "Surgery." "Medicine" refers to the practice of non-operative medicine, and most of its subspecialties require preliminary training in Internal Medicine. In the UK, this was traditionally evidenced by passing the examination for the Membership of the Royal College of Physicians (MRCP) or the equivalent college in Scotland or Ireland. "Surgery" refers to the practice of operative medicine, and most subspecialties in this area require preliminary training in General Surgery, which in the UK leads to membership of the Royal College of Surgeons of England (MRCS). At present, some specialties of medicine do not fit easily into either of these categories, such as radiology, pathology, or anesthesia. Most of these have branched from one or other of the two camps above; for example anaesthesia developed first as a faculty of the Royal College of Surgeons (for which MRCS/FRCS would have been required) before becoming the Royal College of Anaesthetists and membership of the college is attained by sitting for the examination of the Fellowship of the Royal College of Anesthetists (FRCA).

Surgical Specialty

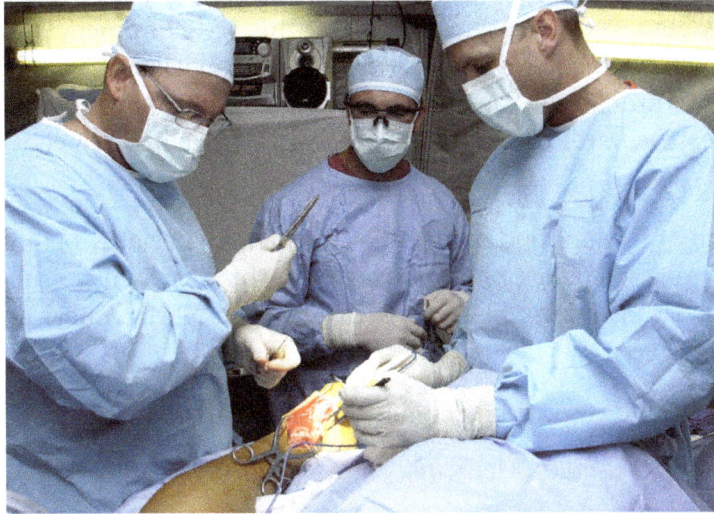

Surgeons in an operating room

Surgery is an ancient medical specialty that uses operative manual and instrumental techniques on a patient to investigate and/or treat a pathological condition such as disease or injury, to help improve bodily function or appearance or to repair unwanted ruptured areas (for example, a perforated ear drum). Surgeons must also manage pre-operative, post-operative, and potential surgical candidates on the hospital wards. Surgery has many sub-specialties, including *general surgery, ophthalmic surgery, cardiovascular surgery, colorectal surgery, neurosurgery, oral and maxillofacial surgery, oncologic surgery, orthopedic surgery, otolaryngology, plastic surgery, podiatric surgery, transplant surgery, trauma surgery, urology, vascular surgery, and pediatric surgery.* In some centers, anesthesiology is part of the division of surgery (for historical and logistical reasons), although it is not a surgical discipline. Other medical specialties may employ surgical procedures, such as ophthalmology and dermatology, but are not considered surgical sub-specialties per se.

Surgical training in the U.S. requires a minimum of five years of residency after medical school. Sub-specialties of surgery often require seven or more years. In addition, fellowships can last an additional one to three years. Because post-residency fellowships can be competitive, many trainees devote two additional years to research. Thus in some cases surgical training will not finish until more than a decade after medical school. Furthermore, surgical training can be very difficult and time-consuming.

Internal Specialty

Internal medicine is the medical specialty dealing with the prevention, diagnosis, and treatment of adult diseases. According to some sources, an emphasis on internal structures is implied. In North America, specialists in internal medicine are commonly called "internists." Elsewhere, especially in Commonwealth nations, such specialists are often called physicians. These terms, *internist* or *physician* (in the narrow sense, common outside North America), generally exclude practitioners of gynecology and obstetrics, pathology, psychiatry, and especially surgery and its subspecialities.

Because their patients are often seriously ill or require complex investigations, internists do much of their work in hospitals. Formerly, many internists were not subspecialized; such *general physicians* would see any complex nonsurgical problem; this style of practice has become much less common. In modern urban practice, most internists are subspecialists: that is, they generally limit their medical practice to problems of one organ system or to one particular area of medical knowledge. For example, gastroenterologists and nephrologists specialize respectively in diseases of the gut and the kidneys.

In the Commonwealth of Nations and some other countries, specialist pediatricians and geriatricians are also described as *specialist physicians* (or internists) who have subspecialized by age of patient rather than by organ system. Elsewhere, especially in North America, general pediatrics is often a form of Primary care.

There are many subspecialities (or subdisciplines) of internal medicine:

- *Angiology/Vascular Medicine*
- *Cardiology*
- *Critical care medicine*
- *Endocrinology*
- *Gastroenterology*
- *Geriatrics*
- *Hematology*
- *Hepatology*
- *Infectious disease*
- *Nephrology*
- *Neurology*
- *Oncology*
- *Pediatrics*
- *Pulmonology/Pneumology/Respirology/chest medicine*
- *Rheumatology*
- Sports Medicine

Training in internal medicine (as opposed to surgical training), varies considerably across the world: In North America, it requires at least three years of residency training after medical school, which can then be followed by a one- to three-year fellowship in the subspecialties listed above. In general, resident work hours in medicine are less than those in surgery, averaging about 60 hours per week in the USA. This difference does not apply in the UK where all doctors are now required by law to work less than 48 hours per week on average.

Diagnostic Specialties

- *Clinical laboratory sciences* are the clinical diagnostic services that apply laboratory techniques to diagnosis and management of patients. In the United States, these services are supervised by a pathologist. The personnel that work in these medical laboratory departments are technically trained staff who do not hold medical degrees, but who usually hold an undergraduate medical technology degree, who actually perform the tests, assays, and procedures needed for providing the specific services. Subspecialties include transfusion medicine, cellular pathology, clinical chemistry, hematology, clinical microbiology and clinical immunology.

- *Pathology as a medical specialty* is the branch of medicine that deals with the study of diseases and the morphologic, physiologic changes produced by them. As a diagnostic specialty, pathology can be considered the basis of modern scientific medical knowledge and plays a large role in evidence-based medicine. Many modern molecular tests such as flow cytometry, polymerase chain reaction (PCR), immunohistochemistry, cytogenetics, gene rearrangements studies and fluorescent in situ hybridization (FISH) fall within the territory of pathology.

- *Diagnostic radiology* is concerned with imaging of the body, e.g. by x-rays, x-ray computed tomography, ultrasonography, and nuclear magnetic resonance tomography. Interventional radiologists can access areas in the body under imaging for an intervention or diagnostic sampling.

- *Nuclear medicine* is concerned with studying human organ systems by administering radiolabelled substances (radiopharmaceuticals) to the body, which can then be imaged outside the body by a gamma camera or a PET scanner. Each radiopharmaceutical consists of two parts: a tracer that is specific for the function under study (e.g., neurotransmitter pathway, metabolic pathway, blood flow, or other), and a radionuclide (usually either a gamma-emitter or a positron emitter). There is a degree of overlap between nuclear medicine and radiology, as evidenced by the emergence of combined devices such as the PET/CT scanner.

- *Clinical neurophysiology* is concerned with testing the physiology or function of the central and peripheral aspects of the nervous system. These kinds of tests can be divided into recordings of: (1) spontaneous or continuously running electrical activity, or (2) stimulus evoked responses. Subspecialties include electroencephalography, electromyography, evoked potential, nerve conduction study and polysomnography. Sometimes these tests are performed by techs without a medical degree, but the interpretation of these tests is done by a medical professional.

Other Major Specialties

The followings are some major medical specialties that do not directly fit into any of the above-mentioned groups.

- *Anesthesiology* (also known as *anaesthetics*): concerned with the perioperative management of the surgical patient. The anesthesiologist's role during surgery is to prevent derangement in the vital organs' (i.e. brain, heart, kidneys) functions and postoperative pain. Outside of the operating room, the anesthesiology physician also serves the same function in the labor & delivery ward, and some are specialized in critical medicine.

- *Dermatology* is concerned with the skin and its diseases. In the UK, dermatology is a subspecialty of general medicine.

- *Emergency medicine* is concerned with the diagnosis and treatment of acute or life-threatening conditions, including trauma, surgical, medical, pediatric, and psychiatric emergencies.

- *Family medicine*, *family practice*, *general practice* or *primary care* is, in many countries, the first port-of-call for patients with non-emergency medical problems. Family physicians often provide services across a broad range of settings including office based practices, emergency room coverage, inpatient care, and nursing home care.

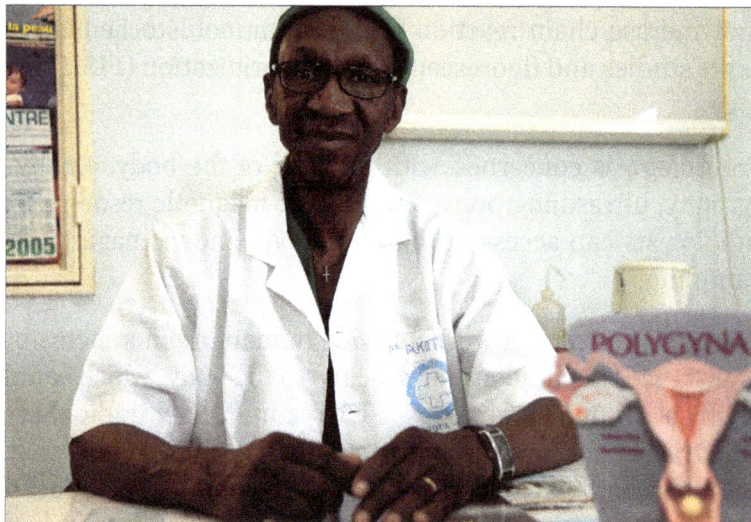

Gynecologist Michel Akotionga of Ouagadougou, Burkina Faso

- *Obstetrics and gynecology* (often abbreviated as *OB/GYN* (American English) or *Obs & Gynae* (British English)) are concerned respectively with childbirth and the female reproductive and associated organs. Reproductive medicine and fertility medicine are generally practiced by gynecological specialists.

- *Medical Genetics* is concerned with the diagnosis and management of hereditary disorders.

- *Neurology* is concerned with diseases of the nervous system. In the UK, neurology is a subspecialty of general medicine.

- *Ophthalmology* is exclusively concerned with the eye and ocular adnexa, combining conservative and surgical therapy.

- *Pediatrics* (AE) or *paediatrics* (BE) is devoted to the care of infants, children, and adolescents. Like internal medicine, there are many pediatric subspecialties for specific age ranges, organ systems, disease classes, and sites of care delivery.

- *Pharmaceutical medicine* is the medical scientific discipline concerned with the discovery, development, evaluation, registration, monitoring and medical aspects of marketing of medicines for the benefit of patients and public health.

- *Physical medicine and rehabilitation* (or *physiatry*) is concerned with functional improvement after injury, illness, or congenital disorders.

- *Podiatric medicine* is the study of, diagnosis, and medical & surgical treatment of disorders of the foot, ankle, lower limb, hip and lower back.

- *Psychiatry* is the branch of medicine concerned with the bio-psycho-social study of the etiology, diagnosis, treatment and prevention of cognitive, perceptual, emotional and behavioral disorders. Related non-medical fields include psychotherapy and clinical psychology.

- *Preventive medicine* is the branch of medicine concerned with preventing disease.

  o *Community health* or *public health* is an aspect of health services concerned with threats to the overall health of a community based on population health analysis.

Interdisciplinary Fields

Some interdisciplinary sub-specialties of medicine include:

- *Aerospace medicine* deals with medical problems related to flying and space travel.

- *Addiction medicine* deals with the treatment of addiction.

- *Medical ethics* deals with ethical and moral principles that apply values and judgments to the practice of medicine.

- *Biomedical Engineering* is a field dealing with the application of engineering principles to medical practice.

- *Clinical pharmacology* is concerned with how systems of therapeutics interact with patients.

- *Conservation medicine* studies the relationship between human and animal health, and environmental conditions. Also known as ecological medicine, environmental medicine, or medical geology.

- *Disaster medicine* deals with medical aspects of emergency preparedness, disaster mitigation and management.

- *Diving medicine* (or hyperbaric medicine) is the prevention and treatment of diving-related problems.

- *Evolutionary medicine* is a perspective on medicine derived through applying evolutionary theory.

- *Forensic medicine* deals with medical questions in legal context, such as determination of the time and cause of death, type of weapon used to inflict trauma, reconstruction of the facial features using remains of deceased (skull) thus aiding identification.

- *Gender-based medicine* studies the biological and physiological differences between the human sexes and how that affects differences in disease.

- *Hospice and Palliative Medicine* is a relatively modern branch of clinical medicine that deals with pain and symptom relief and emotional support in patients with terminal illnesses including cancer and heart failure.

- *Hospital medicine* is the general medical care of hospitalized patients. Physicians whose primary professional focus is hospital medicine are called hospitalists in the USA and Canada. The term Most Responsible Physician (MRP) or attending physician is also used interchangeably to describe this role.

- *Laser medicine* involves the use of lasers in the diagnostics and/or treatment of various conditions.

- *Medical humanities* includes the humanities (literature, philosophy, ethics, history and religion), social science (anthropology, cultural studies, psychology, sociology), and the arts (literature, theater, film, and visual arts) and their application to medical education and practice.

- *Health informatics* is a relatively recent field that deal with the application of computers and information technology to medicine.

- *Nosology* is the classification of diseases for various purposes.

- *Nosokinetics* is the science/subject of measuring and modelling the process of care in health and social care systems.

- *Occupational medicine*'s principal role is the provision of health advice to organizations and individuals to ensure that the highest standards of health and safety at work can be achieved and maintained.

- *Pain management* (also called *pain medicine*, or *algiatry*) is the medical discipline concerned with the relief of pain.

- *Pharmacogenomics* is a form of *individualized medicine.*

- *Podiatric medicine* is the study of, diagnosis, and medical treatment of disorders of the foot, ankle, lower limb, hip and lower back.

- *Sexual medicine* is concerned with diagnosing, assessing and treating all disorders related to sexuality.

- *Sports medicine* deals with the treatment and prevention and rehabilitation of sports/exercise injuries such as muscle spasms, muscle tears, injuries to ligaments (ligament tears or ruptures) and their repair in athletes, amateur and professional.

- *Therapeutics* is the field, more commonly referenced in earlier periods of history, of the various remedies that can be used to treat disease and promote health.

- *Travel medicine* or *emporiatrics* deals with health problems of international travelers or travelers across highly different environments.

- *Tropical medicine* deals with the prevention and treatment of tropical diseases. It is studied separately in temperate climates where those diseases are quite unfamiliar to medical practitioners and their local clinical needs.

- *Urgent care* focuses on delivery of unscheduled, walk-in care outside of the hospital emergency department for injuries and illnesses that are not severe enough to require care in an emergency department. In some jurisdictions this function is combined with the emergency room.

- Veterinary medicine; veterinarians apply similar techniques as physicians to the care of animals.

- *Wilderness medicine* entails the practice of medicine in the wild, where conventional medical facilities may not be available.

- Many other health science fields, e.g. dietetics

Education and Legal Controls

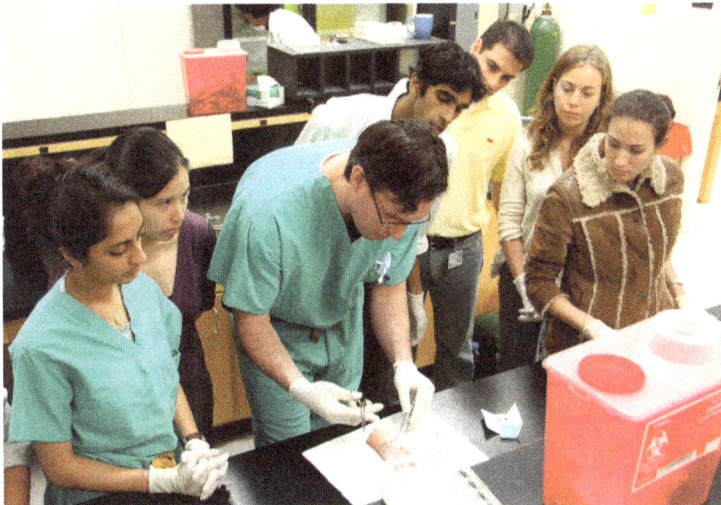

Medical students learning about stitches

Medical education and training varies around the world. It typically involves entry level education at a university medical school, followed by a period of supervised practice or internship, and/or residency. This can be followed by postgraduate vocational training. A variety of teaching methods have been employed in medical education, still itself a focus of active research. In Canada and the United States of America, a Doctor of Medicine degree, often abbreviated M.D., or a Doctor of Osteopathic Medicine degree, often abbreviated as D.O. and unique to the United States, must be completed in and delivered from a recognized university.

Since knowledge, techniques, and medical technology continue to evolve at a rapid rate, many regulatory authorities require continuing medical education. Medical practitioners upgrade their knowledge in various ways, including medical journals, seminars, conferences, and online programs.

In most countries, it is a legal requirement for a medical doctor to be licensed or registered. In general, this entails a medical degree from a university and accreditation by a medical board or an equivalent national organization, which may ask the applicant to pass exams. This restricts the considerable legal authority of the medical profession to physicians that are trained and qualified by national standards. It is also intended as an assurance to patients and as a safeguard against charlatans that practice inadequate medicine for personal gain. While the laws generally require medical doctors to be trained in "evidence based", Western, or Hippocratic Medicine, they are not intended to discourage different paradigms of health.

Headquarters of the Organización Médica Colegial de España, which regulates the medical profession in Spain

In the European Union, the profession of doctor of medicine is regulated. A profession is said to be regulated when access and exercise is subject to the possession of a specific professional qualification. The regulated professions database contains a list of regulated professions for doctor of medicine in the EU member states, EEA countries and Switzerland. This list is covered by the Directive 2005/36/EC.

Doctors who are negligent or intentionally harmful in their care of patients can face charges of medical malpractice and be subject to civil, criminal, or professional sanctions.

Medical Ethics

Medical ethics is a system of moral principles that apply values and judgments to the practice of medicine. As a scholarly discipline, medical ethics encompasses its practical application in clinical settings as well as work on its history, philosophy, theology, and sociology. Six of the values that commonly apply to medical ethics discussions are:

- autonomy - the patient has the right to refuse or choose their treatment. (*Voluntas aegroti suprema lex.*)

- beneficence - a practitioner should act in the best interest of the patient. (*Salus aegroti suprema lex.*)

- justice - concerns the distribution of scarce health resources, and the decision of who gets what treatment (fairness and equality).

- non-maleficence - "first, do no harm" (*primum non-nocere*).

- respect for persons - the patient (and the person treating the patient) have the right to be treated with dignity.

- truthfulness and honesty - the concept of informed consent has increased in importance since the historical events of the Doctors' Trial of the Nuremberg trials, Tuskegee syphilis experiment, and others.

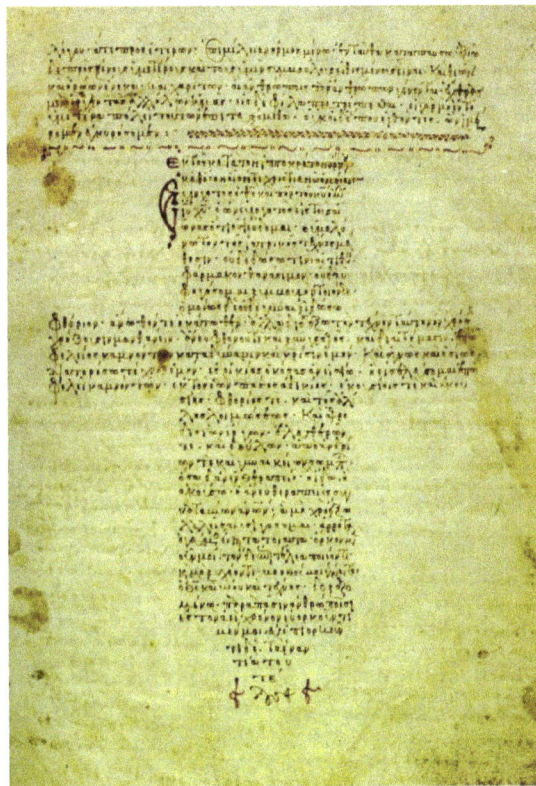

A 12th-century Byzantine manuscript of the Hippocratic Oath

Values such as these do not give answers as to how to handle a particular situation, but provide a useful framework for understanding conflicts. When moral values are in conflict, the result may be an ethical dilemma or crisis. Sometimes, no good solution to a dilemma in medical ethics exists, and occasionally, the values of the medical community (i.e., the hospital and its staff) conflict with the values of the individual patient, family, or larger non-medical community. Conflicts can also arise between health care providers, or among family members. For example, some argue that the principles of autonomy and beneficence clash when patients refuse blood transfusions, considering them life-saving; and truth-telling was not emphasized to a large extent before the HIV era.

History

Statuette of ancient Egyptian physician Imhotep, the first physician from antiquity known by name.

Ancient World

Prehistoric medicine incorporated plants (herbalism), animal parts, and minerals. In many cases these materials were used ritually as magical substances by priests, shamans, or medicine men. Well-known spiritual systems include animism (the notion of inanimate objects having spirits), spiritualism (an appeal to gods or communion with ancestor spirits); shamanism (the vesting of an individual with mystic powers); and divination (magically obtaining the truth). The field of medical anthropology examines the ways in which culture and society are organized around or impacted by issues of health, health care and related issues.

Early records on medicine have been discovered from ancient Egyptian medicine, Babylonian Medicine, Ayurvedic medicine (in the Indian subcontinent), classical Chinese medicine (predecessor to the modern traditional Chinese Medicine), and ancient Greek medicine and Roman medicine.

In Egypt, Imhotep (3rd millennium BC) is the first physician in history known by name. The oldest Egyptian medical text is the *Kahun Gynaecological Papyrus* from around 2000 BCE, which describes gynaecological diseases. The *Edwin Smith Papyrus* dating back to 1600 BCE is an early work on surgery, while the *Ebers Papyrus* dating back to 1500 BCE is akin to a textbook on medicine.

In China, archaeological evidence of medicine in Chinese dates back to the Bronze Age Shang Dynasty, based on seeds for herbalism and tools presumed to have been used for surgery. The *Huangdi Neijing*, the progenitor of Chinese medicine, is a medical text written beginning in the 2nd century BCE and compiled in the 3rd century.

In India, the surgeon Sushruta described numerous surgical operations, including the earliest forms of plastic surgery. Earliest records of dedicated hospitals come from Mihintale in Sri Lanka where evidence of dedicated medicinal treatment facilities for patients are found.

In Greece, the Greek physician Hippocrates, the "father of western medicine", laid the foundation for a rational approach to medicine. Hippocrates introduced the Hippocratic Oath for physicians, which is still relevant and in use today, and was the first to categorize illnesses as acute, chronic, endemic and epidemic, and use terms such as, "exacerbation, relapse, resolution, crisis, paroxysm, peak, and convalescence". The Greek physician Galen was also one of the greatest surgeons of the ancient world and performed many audacious operations, including brain and eye surgeries. After the fall of the Western Roman Empire and the onset of the Early Middle Ages, the Greek tradition of medicine went into decline in Western Europe, although it continued uninterrupted in the Eastern Roman (Byzantine) Empire.

Most of our knowledge of ancient Hebrew medicine during the 1st millennium BC comes from the Torah, i.e. the Five Books of Moses, which contain various health related laws and rituals. The Hebrew contribution to the development of modern medicine started in the Byzantine Era, with the physician Asaph the Jew.

Middle Ages

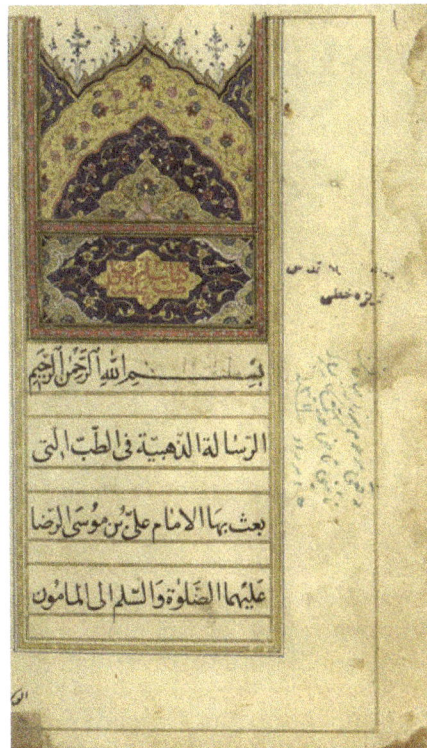

A manuscript of *Al-Risalah al-Dhahabiah* by Ali al-Ridha, the eighth Imam of Shia Muslims. The text says: "Golden dissertation in medicine which is sent by Imam Ali ibn Musa al-Ridha, peace be upon him, to al-Ma'mun."

After 750 CE, the Muslim world had the works of Hippocrates, Galen and Sushruta translated into Arabic, and Islamic physicians engaged in some significant medical research. Notable Islamic medical pioneers include the Persian polymath, Avicenna, who, along with Imhotep and Hippocrates, has also been called the "father of medicine". He wrote *The Canon of Medicine*, considered one of the most famous books in the history of medicine. Others include Abulcasis, Avenzoar, Ibn al-Nafis, and Averroes. Rhazes was one of the first to question the Greek theory of humorism, which nevertheless remained influential in both medieval Western and medieval Islamic medicine. *Al-Risalah al-Dhahabiah* by Ali al-Ridha, the eighth Imam of Shia Muslims, is revered as the most precious Islamic literature in the Science of Medicine. The Islamic Bimaristan hospitals were an early example of public hospitals.

In Europe, Charlemagne decreed that a hospital should be attached to each cathedral and monastery and the historian Geoffrey Blainey likened the activities of the Catholic Church in health care during the Middle Ages to an early version of a welfare state: "It conducted hospitals for the old and orphanages for the young; hospices for the sick of all ages; places for the lepers; and hostels or inns where pilgrims could buy a cheap bed and meal". It supplied food to the population during famine and distributed food to the poor. This welfare system the church funded through collecting taxes on a large scale and possessing large farmlands and estates. The Benedictine order was noted for setting up hospitals and infirmaries in their monasteries, growing medical herbs and becoming the chief medical care givers of their districts, as at the great Abbey of Cluny. The Church also established a network of cathedral schools and universities where medicine was studied. The Schola Medica Salernitana in Salerno, looking to the learning of Greek and Arab physicians, grew to be the finest medical school in Medieval Europe.

Panorama of Siena's Santa Maria della Scala Hospital, one of Europe's oldest hospitals. During the Middle Ages, the Catholic Church established universities which revived the study of sciences - drawing on the learning of Greek and Arab physicians in the study of medicine.

However, the fourteenth and fifteenth century Black Death devastated both the Middle East and Europe, and it has even been argued that Western Europe was generally more effective in recovering from the pandemic than the Middle East. In the early modern period, important early figures in medicine and anatomy emerged in Europe, including Gabriele Falloppio and William Harvey.

The major shift in medical thinking was the gradual rejection, especially during the Black Death in the 14th and 15th centuries, of what may be called the 'traditional authority' approach to science and medicine. This was the notion that because some prominent person in the past said something must be so, then that was the way it was, and anything one observed to the contrary was an anomaly (which was paralleled by a similar shift in European society in general). Physicians like Vesalius improved upon or disproved some of the theories from the past. The main tomes used both by medicine students and expert physicians were Materia Medica and Pharmacopoeia.

Andreas Vesalius was the author of *De humani corporis fabrica*, an important book on human anatomy. Bacteria and microorganisms were first observed with a microscope by Antonie van Leeuwenhoek in 1676, initiating the scientific field microbiology. Independently from Ibn al-Nafis, Michael Servetus rediscovered the pulmonary circulation, but this discovery did not reach the public because it was written down for the first time in the "Manuscript of Paris" in 1546, and later published in the theological work for which he paid with his life in 1553. Later this was described by Renaldus Columbus and Andrea Cesalpino. Herman Boerhaave is sometimes referred to as a "father of physiology" due to his exemplary teaching in Leiden and textbook 'Institutiones medicae' (1708). Pierre Fauchard has been called "the father of modern dentistry".

Modern

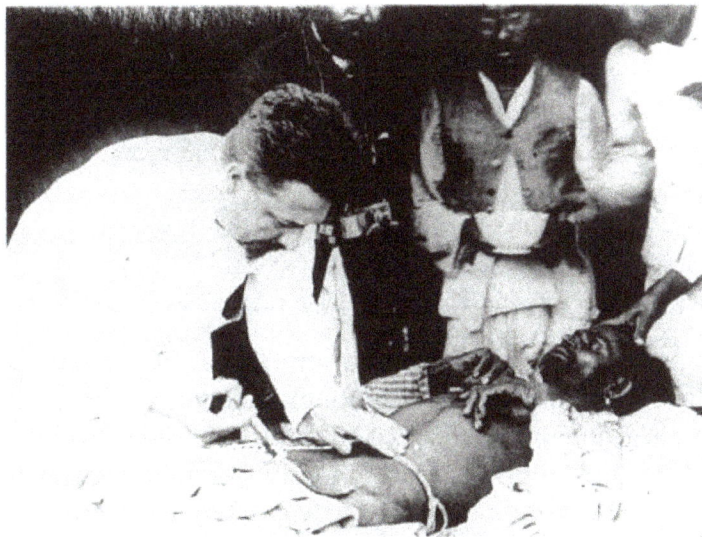

Paul-Louis Simond injecting a plague vaccine in Karachi, 1898

Veterinary medicine was, for the first time, truly separated from human medicine in 1761, when the French veterinarian Claude Bourgelat founded the world's first veterinary school in Lyon, France. Before this, medical doctors treated both humans and other animals.

Modern scientific biomedical research (where results are testable and reproducible) began to replace early Western traditions based on herbalism, the Greek "four humours" and other such pre-modern notions. The modern era really began with Edward Jenner's discovery of the smallpox vaccine at the end of the 18th century (inspired by the method of inoculation earlier practiced in Asia), Robert Koch's discoveries around 1880 of the transmission of disease by bacteria, and then the discovery of antibiotics around 1900.

The post-18th century modernity period brought more groundbreaking researchers from Europe. From Germany and Austria, doctors Rudolf Virchow, Wilhelm Conrad Röntgen, Karl Landsteiner and Otto Loewi made notable contributions. In the United Kingdom, Alexander Fleming, Joseph Lister, Francis Crick and Florence Nightingale are considered important. Spanish doctor Santiago Ramón y Cajal is considered the father of modern neuroscience.

From New Zealand and Australia came Maurice Wilkins, Howard Florey, and Frank Macfarlane Burnet.

In the United States, William Williams Keen, William Coley, James D. Watson, Italy (Salvador Luria), Switzerland (Alexandre Yersin), Japan (Kitasato Shibasaburō), and France (Jean-Martin Charcot, Claude Bernard, Paul Broca) and others did significant work. Russian Nikolai Korotkov also did significant work, as did Sir William Osler and Harvey Cushing.

Alexander Fleming's discovery of penicillin in September 1928 marks the start of modern antibiotics.

As science and technology developed, medicine became more reliant upon medications. Throughout history and in Europe right until the late 18th century, not only animal and plant products were used as medicine, but also human body parts and fluids. Pharmacology developed in part from herbalism and some drugs are still derived from plants (atropine, ephedrine, warfarin, aspirin, digoxin, vinca alkaloids, taxol, hyoscine, etc.). Vaccines were discovered by Edward Jenner and Louis Pasteur.

The first antibiotic was arsphenamine (Salvarsan) discovered by Paul Ehrlich in 1908 after he observed that bacteria took up toxic dyes that human cells did not. The first major class of antibiotics was the sulfa drugs, derived by German chemists originally from azo dyes.

Pharmacology has become increasingly sophisticated; modern biotechnology allows drugs targeted towards specific physiological processes to be developed, sometimes designed for compatibility with the body to reduce side-effects. Genomics and knowledge of human genetics is having some influence on medicine, as the causative genes of most monogenic genetic disorders have now been identified, and the development of techniques in molecular biology and genetics are influencing medical technology, practice and decision-making.

Evidence-based medicine is a contemporary movement to establish the most effective algorithms of practice (ways of doing things) through the use of systematic reviews and meta-analysis. The movement is facilitated by modern global information science, which allows as much of the available evidence as possible to be collected and analyzed according to standard protocols that are then disseminated to healthcare providers. The Cochrane Collaboration leads this movement. A

2001 review of 160 Cochrane systematic reviews revealed that, according to two readers, 21.3% of the reviews concluded insufficient evidence, 20% concluded evidence of no effect, and 22.5% concluded positive effect.

Traditional Medicine

Traditional medicine (also known as indigenous or folk medicine) comprises knowledge systems that developed over generations within various societies before the era of modern medicine. The World Health Organization (WHO) defines traditional medicine as "the sum total of the knowledge, skills, and practices based on the theories, beliefs, and experiences indigenous to different cultures, whether explicable or not, used in the maintenance of health as well as in the prevention, diagnosis, improvement or treatment of physical and mental illness."

In some Asian and African countries, up to 80% of the population relies on traditional medicine for their primary health care needs. When adopted outside of its traditional culture, traditional medicine is often called alternative medicine. Practices known as traditional medicines include Ayurveda, Siddha medicine, Unani, ancient Iranian medicine, Irani, Islamic medicine, traditional Chinese medicine, traditional Korean medicine, acupuncture, Muti, Ifá, and traditional African medicine.

The WHO notes however that "inappropriate use of traditional medicines or practices can have negative or dangerous effects" and that "further research is needed to ascertain the efficacy and safety" of several of the practices and medicinal plants used by traditional medicine systems. The line between alternative medicine and quackery is a contentious subject.

Traditional medicine may include formalized aspects of folk medicine, that is to say longstanding remedies passed on and practised by lay people. Folk medicine consists of the healing practices and ideas of body physiology and health preservation known to some in a culture, transmitted informally as general knowledge, and practiced or applied by anyone in the culture having prior experience. Folk medicine may also be referred to as traditional medicine, alternative medicine, indigenous medicine, or natural medicine. These terms are often considered interchangeable, even though some authors may prefer one or the other because of certain overtones they may be willing to highlight. In fact, out of these terms perhaps only *indigenous medicine* and *traditional medicine* have the same meaning *folk medicine*, while the others should be understood rather in a modern or modernized context.

References

- Committee on Forestry Research, National Research Council (1990). Forestry Research: A Mandate for Change. National Academies Press. pp. 15–16. ISBN 0-309-04248-8.

- Budowski, Gerardo (1982). "Applicability of agro-forestry systems". In MacDonald, L.H. Agro-forestry in the African Humid Tropics. United Nations University. ISBN 92-808-0364-6. Retrieved 17 March 2016.

- Broudy, Eric (1979). The Book of Looms: A History of the Handloom from Ancient Times to the Present. UPNE. p. 81. ISBN 978-0-87451-649-4.

- Berg, Paul; Singer, Maxine (15 August 2003). George Beadle: An Uncommon Farmer. The Emergence of Genetics in the 20th century. Cold Springs Harbor Laboratory Press. ISBN 978-0-87969-688-7.

- Coulehan JL, Block MR (2005). The Medical Interview: Mastering Skills for Clinical Practice (5th ed.). F. A. Davis. ISBN 0-8036-1246-X. OCLC 232304023.

- Fowler, H.W. (1994). A Dictionary of Modern English Usage (Wordsworth Collection) (Wordsworth Collection). NTC/Contemporary Publishing Company. ISBN 1-85326-318-4.

- Unschuld, Pual (2003). Huang Di Nei Jing: Nature, Knowledge, Imagery in an Ancient Chinese Medical Text. University of California Press. p. ix. ISBN 978-0-520-92849-7.

- Peter Barrett (2004). Science and Theology Since Copernicus: The Search for Understanding. Continuum International Publishing Group. p. 18. ISBN 0-567-08969-X.

- Acharya, Deepak and Shrivastava Anshu (2008): Indigenous Herbal Medicines: Tribal Formulations and Traditional Herbal Practices, Aavishkar Publishers Distributor, Jaipur- India. ISBN 978-81-7910-252-7.

- "Africa may be able to feed only 25% of its population by 2025". Mongabay. 14 December 2006. Archived from the original on 27November 2011. Retrieved 15 July 2016.

- Martin Heller; Gregory Keoleian (2000). "Life Cycle-Based Sustainability Indicators for Assessment of the U.S. Food System" (PDF). University of Michigan Center for Sustainable Food Systems. Retrieved 17 March 2016.

- Muhammad Jawad Fadlallah. Imam ar-Ridha', A Historical and Biographical Research. Al-islam.org. Yasin T. Al-Jibouri. Retrieved 18 Jun 2014.

- International Food Policy Research Institute (2014). "Food Security in a World of Growing Natural Resource Scarcity". CropLife International. Retrieved 1 July 2013.

- Molden, D. (ed.). "Findings of the Comprehensive Assessment of Water Management in Agriculture" (PDF). Annual Report 2006/2007. International Water Management Institute. Retrieved 6 January 2014.

- Bai, Z.G.; D.L. Dent; L. Olsson & M.E. Schaepman (November 2008). "Global assessment of land degradation and improvement: 1. identification by remote sensing" (PDF). FAO/ISRIC. Retrieved 24 May 2013.

- Pimentel, D.; T.W. Culliney; T. Bashore (1996). "Public health risks associated with pesticides and natural toxins in foods". Radcliffe's IPM World Textbook. Archived from the original on 2 May 2013. Retrieved 7 May 2013.

- Safefood Consulting, Inc. (2005). "Benefits of Crop Protection Technologies on Canadian Food Production, Nutrition, Economy and the Environment". CropLife International. Retrieved 24 May 2013.

- Runge, C. Ford (June 2006). "Agricultural Economics: A Brief Intellectual History" (PDF). Center for International Food and Agriculture Policy. p. 4. Retrieved 16 September 2013.

- Conrad, David E. "Tenant Farming and Sharecropping". Encyclopedia of Oklahoma History and Culture. Oklahoma Historical Society. Retrieved 16 September 2013.

- Peter J. Lloyd; Johanna L. Croser; Kym Anderson (March 2009). "How Do Agricultural Policy Restrictions to Global Trade and Welfare Differ across Commodities?" (PDF). Policy Research Working Paper #4864. The World Bank. pp. 2–3. Retrieved 16 April 2013.

- Kym Anderson; Ernesto Valenzuela (April 2006). "Do Global Trade Distortions Still Harm Developing Country Farmers?" (PDF). World Bank Policy Research Working Paper 3901. World Bank. pp. 1–2. Retrieved 16 April 2013.

Permissions

Index